Edward R. Scheinerman

MW00951580

Mathematical Notation
A Guide for Engineers and Scientists

Author

Edward R. Scheinerman
Department of Applied Mathematics and Statistics
Johns Hopkins University
Baltimore, Maryland 21218 USA
http://www.ams.jhu.edu/~ers

Cover illustration and design

Jonah Scheinerman
http://www.scheinerman.net/jonah

Library of Congress Control Number: 2011914841
ISBN-13: 978-1466230521
ISBN-10: 1466230525
2011:09:20:13:24

To Nick and Andrew

*In mathematics you don't understand things.
You just get used to them.*

—John von Neumann

Contents

Preface

Mathematics is the language of science and engineering. Like all languages, it has its own special vocabulary. When confronted with an unfamiliar *word* from its lexicon, we can search for that term in our textbooks or on the web using tools such as Google or Mathworld [**13**]. But what do we do when confounded by some strange arrow or squiggle? Just how would one type $\mapsto$ or $\ni$ into a search engine?

The goal of this *Guide* is to provide a solution to this conundrum. We have collected a wide assortment of mathematical symbols and notation commonly[1] used in science and engineering for your easy reference.

How to use this *Guide*

Our aim is twofold. First, we provide a tool for readers to look up notation they may have confronted in a class or a textbook. Here's how you'll locate it in this *Guide*.

- If the unknown notation you seek is based on a Latin letter such as $\forall$ or $\mathcal{L}$, or a word such as det, then search the Alphabetical Notation Index (on page 73) under the appropriate letter (A, L, and D in these examples).
- If the unknown notation is based on a Greek letter, then search the Greek Notation Index (page 77). The Greek alphabet is shown in Figure 1.1 on page 2.
- If the notation is some other sort of symbol (a decoration on a letter, an operation symbol, an arrow, and so on), then consult the Chart (page 69). The Chart also provides LaTeX code[2] for typesetting the notation.
- If you know the general area of mathematics from which the notation arises, then you can proceed directly to the relevant chapter for that discipline. Alternatively, you can consult the Topic Index (page 79).

---

[1]We do not claim to cover all notation used by mathematicians; see [**11**] to understand why that task would be overwhelming. We focus on notation one is likely to see in university-level courses.

[2]This is useful only for readers with some familiarity with LaTeX. We strongly encourage the use of LaTeX for technical writing. For good introductions see [**2**], [**5**], or [**6**]. For advanced information, refer to [**3**].

Second, we provide guidance on how to use notation. For example, what symbol should you use for a vector of all 1s? (Answer on page 39.) When there is more than one choice (e.g., $f'(x)$, dy/dx, or $\dot{y}$ for derivative) we show the various forms, and sometimes express a preference for one notation versus another.

How *not* to use this *Guide*

Do not expect to learn new mathematics from this book; that's not our goal. Rather, enough information is provided to *remind* you about notation/ideas you may have previously encountered or to search for an appropriate reference for full detail.

On occasion we use notation in one chapter before it is presented in a later chapter. This is consistent with our philosophy that we are trying either to assist readers with notation for concepts they already know or to provide sufficient context so they can consult an appropriate reference.

This *Guide* does not address discipline specific notation. For example, chemists use square brackets to denote concentration $[Na^+]$, but this is a convention from chemistry, not mathematics, and so it is beyond the scope of this *Guide*. We also do not cover notation for physical constants such as $\hbar$ (Planck's constant divided by 2π).

We hope you enjoy your journey through the jungle of mathematical notation and that you quickly find where $\otimes$ marks the spot.

Acknowledgments

I greatly appreciate the ideas, feedback, and corrections I received from colleagues and students. Many thanks to: Greg Chirikjian, Paul Dagdigian, Lara Diamond, Tak Igusa, Dan Naiman, Christine Nickel, Jessica Noviello, Daniel Scheinerman, Avik Som, Jim Spicer, and Alex Strachan.

Special thanks to Jonah Scheinerman for extensive stylistic and editorial help as well as for the art and design of the cover of this book.

I welcome your feedback and comments. Please visit the web page for this book

```
http://www.ams.jhu.edu/~ers/notation
```

for updates and errata.

—Ed Scheinerman, Baltimore

CHAPTER 1

Letters

1. The Latin alphabet

In mathematics we use letters to name mathematical objects. These objects are often numbers, but they may also be functions, vectors, matrices, sets, and so forth.

The familiar Latin alphabet is the most used (and for many students, the most comfortable) set of letters, appearing in upper and lower case versions. In most mathematics books, these letters are typeset in an italic font, like this: $x + y - 3$. Notice that the numeral 3 is not set in italics. However, when written by hand, most people do not attempt to distinguish between roman x and italic x. Some books use roman letters for well-known mathematical constants such as e ≈ 2.718 and i $= \sqrt{-1}$.

Named functions (such as those from trigonometry) are written as short words: $\cos(x)$, $\log(z + 2)$, $\det(A)$. Notice that these are written in roman.

Upper and lower case letters are considered different. Thus $X + x$ is not equivalent to $2x$.

One may also see different type styles applied to letters, such as bold **x** and script $\mathcal{X}$. Bold letters are often used to denote vectors. Bold letters can be difficult to draw on the board in class; instead, various accents and decorations can be used. For example, instead of writing **x**, one may write $\vec{x}$ to denote that x is a vector. A wavy underline also indicates bold: x. However, writing a plain x to stand for a vector is not unusual.

Another way in which bold letters are written by hand (and also in print) is to double some portion of the shape of the letter. For example, in lieu of **Z**, one writes $\mathbb{Z}$. This style is called *blackboard bold* and is typically reserved for specific sets of numbers such as the real numbers $\mathbb{R}$ or the complex numbers $\mathbb{C}$.

The use of different styles of letters is especially useful to distinguish different sorts of mathematical objects. For example, in linear algebra it is helpful to use italic lower case letters to denote numbers (scalars), bold lower case letters to denote vectors, and upper case letters to denote matrices.

2. The Greek alphabet

Somehow the 52 upper and lower case Latin letters, in various font styles (italic, bold, script) are just not sufficient, and for this reason letters from other alphabets are brought into service. The most common choice after the Latin alphabet is the Greek alphabet. The letters of the Greek alphabet are presented in Figure 1.1.

Name	Lower case	Upper case
Alpha	α	A
Beta	β	B
Gamma	γ	Γ
Delta	δ	Δ
Epsilon	ϵ or ε	E
Zeta	ζ	Z
Eta	η	H
Theta	θ or ϑ	Θ
Iota	ι	I
Kappa	κ	K
Lambda	λ	Λ
Mu	μ	M
Nu	ν	N
Xi	ξ	Ξ
Omicron	o	O
Pi	π	Π
Rho	ρ	R
Sigma	σ	Σ
Tau	τ	T
Upsilon	υ	Υ
Phi	ϕ or φ	Φ
Chi	χ	X
Psi	ψ	Ψ
Omega	ω	Ω

FIGURE 1.1. The Greek alphabet.

Some Greek letters look exactly the same as their Latin counterparts (e.g., upper case mu and em) and therefore cannot be used to stand for different things.

Letters from other alphabets (such as German ℌ or Hebrew ℵ) are used in mathematics, but these are not often seen in science and engineering settings.

3. Decorations

The repertoire provided by the dozens of letters available from the Latin and Greek alphabets (in their various cases and type styles) is often expanded by various decorations attached to these letters. As we noted earlier, one often places a small arrow above a letter to denote a vector: $\vec{v}$.

The most common decorations attached to letters are subscripts. For example, if one is referring to several points in the plane, it is natural to designate these points by coordinates with numerical subscripts:

$$(x_1, y_1), \ (x_2, y_2), \ \ldots, \ (x_9, y_9)$$

Superscripts are usually reserved for exponentiation; that is, x^2 means x-squared. But sometimes superscripts are also used to denote members of a series; see, for example, the discussion on page 47 about tensors.

The prime mark ($\prime$) attached to letters can be used to denote a different object. For example, x and x' might be two different numbers. One may place multiple prime marks on a letter to create several different names: a, a', a'', a'''. This can easily become unreadable after three marks. In that case, some people write small (lower case) roman numerals to show the number of primes like this: $a^{(\text{iv})}$ in place of a''''. In calculus, the prime mark is reserved for differentiation.

Drawing a wavy underline beneath a letter is a copy editor's convention for indicating that a certain bit of text is to be set in boldface, and occasionally this notation is used in handwritten mathematics: $\underline{x}$.

Other decorations are attached to letters to create additional names. Some that appear often are the hat ($\hat{x}$), the bar ($\bar{x}$), the dot ($\dot{x}$), and multiple dots ($\ddot{x}$).

4. Traditional uses

In principle, any letter can be used to stand for any mathematical object. However, in certain disciplines there are strong traditions to reserve certain letters for specific purposes. For example, in much of science and engineering the Greek letter π stands for the familiar ratio of a circle's circumference to its diameter, approximately 3.14159. However, in other branches of mathematics, the symbol π can have other meanings. In general, of course, it is best to use symbols in a discipline's conventional manner.

CHAPTER 2

Collections

1. Sets

A *set* is an unordered collection of things in which repetition is forbidden. The simplest way to specify a set is to list its objects between curly braces: $\{1, 2, 7\}$. This is a set containing exactly three objects: the numbers 1, 2, and 7. Order and repetition do not matter, so $\{2, 1, 7\}$ and $\{7, 1, 1, 2\}$ are the same set with precisely the same three elements.

Membership in a set is indicated with the special symbol $\in$ which is often pronounce "is an element of". The symbol looks like the Greek epsilon (ϵ) but is, in fact, different (though some people use it for set membership nonetheless). Thus $x \in A$ asserts that the object x is an element of the set A. For example: $2 \in \{1, 2, 7\}$. The symbol $\notin$ means *not an element of* as in $3 \notin \{1, 2, 7\}$. When the symbol $\in$ is written backwards[1] ($\ni$) it means, *includes the element*, like this: $A \ni x$. This means exactly the same thing as $x \in A$. For example, $\{1, 2, 7\} \ni 7$.

The set containing no elements is called the *empty set* or *null set*. It is denoted by the symbol $\emptyset$ or $\varnothing$, or by an empty pair of braces { }. Of these, $\varnothing$ is the clearest.

Large sets may be specified by listing a few elements and then suggesting the pattern continues using ellipses ($\dots$). This is handy for both finite and infinite sets provided the pattern established is crystal clear. For example: $\{1, 2, 3, \dots, 100\}$ is clearly the set of integers from 1 to 100 inclusive and $\{2, 4, 6, 8, \dots\}$ is the set of positive even numbers.

However, a less ambiguous (if more elaborate) way to describe a set is to use *set builder notation*. The notation looks like this:

$$\{x \in \mathbb{Z} : 1 \le x \le 100\}.$$

The letter x is a *dummy variable*; its purpose is to stand for a generic name for the elements of this set. Next we read "$\in \mathbb{Z}$" which means the elements of this set come from the set of integers[2]. After this preamble, which is punctuated by a colon, is a *condition* that tells us *which* integers are in this set. In this case, they are the integers that are at least 1 and at most 100. Incidentally, the set of integers from 1 to n inclusive is sometimes written $[n]$.

---

[1] The symbol $\ni$ (or $\ni$) is sometimes used as an abbreviation for the words *such that*. If one wishes to abbreviate this expression, it is much clearer to simply write "s.t."

[2] The symbol $\mathbb{Z}$ stands for the set of all integers; see §4.2 on page 15.

Sometimes the colon is replaced by a vertical bar | as in $\{u \in \mathbb{Z} \mid u \geq 100\}$. This, of course, is the set $\{100, 101, 102, \ldots\}$.

The condition after the colon or vertical bar may be expressed in words. For example:

$$\{a \in \mathbb{Z} : a \text{ is divisible by } 5\}$$

is the set $\{\ldots, -15, -10, -5, 0, 5, 10, 15, \ldots\}$.

If the context is clear, this notation may be abbreviated to omit the set to the left of the colon; for example, $\{n : n > 5\}$. The problem, of course, is that it is unclear if this means all real numbers that are greater than 5, or just the integers 6 and up.

More elaborate expressions may appear to the left of the colon. For example, $\{(x, y) : x + y = 5\}$ stands for a set of ordered pairs that includes the element $(2, 3)$ but not the element $(4, 4)$.

There are a variety of relations and operations involving sets; here are the ones most frequently encountered.

- **Set equality**. If A and B are sets, then $A = B$ means that A and B have exactly the same elements.
- **Subset**. If A and B are sets, $A \subseteq B$ means that every element of A is also an element of B. Some older books (and professors) might simply write $A \subset B$. It would be preferable to reserve $\subset$ to mean "subset but not equal" so that $\subset$ / $\subseteq$ are analogous to $<$ / $\leq$, but this usage is not universally accepted.

  Note that $A \subseteq B$ does not mean the same thing as $A \in B$.
- **Superset**. $A \supseteq B$ means $B \subseteq A$. That is, the elements of A include all the elements of B (and possibly more). Some people write $\supset$ for $\supseteq$.
- **Union**. If A and B are sets, then $A \cup B$ (the *union* of A and B) is the set whose elements that are in A or B or both. For example,

  $$\{1, 2, 3\} \cup \{2, 4, 6\} = \{1, 2, 3, 4, 6\}.$$

  A variety of special notation is used to indicate the *disjoint union* of sets. Some of the symbols used are $+$, $\oplus$, $\uplus$, $\cup$, $\sqcup$, and $\coprod$. In all cases the notation takes on a double meaning. First, it asserts the sets in question are pairwise disjoint (no element is common to any two); second, it expresses the union of those sets. For example, if you read

  $$A_1 \uplus A_2 \uplus \cdots \uplus A_n = B$$

  then you know (a) that $A_i \cap A_j = \emptyset$ for all $i \neq j$ and (b) the union of the sets $A_1, A_2, \ldots, A_n$ is B.
- **Intersection**. If A and B are sets, then $A \cap B$ (the *intersection* of A and B) is the set whose elements that are in both A and B. For example,

  $$\{1, 2, 3\} \cap \{2, 4, 6\} = \{2\}.$$

- **Difference**. If A and B are sets, then $A - B$ is the set of all elements of A that are not elements of B. For example,

$$\{1, 2, 3\} - \{2, 4, 6\} = \{1, 3\}.$$

Some people write set difference like this $A \setminus B$ to distinguish set difference from ordinary subtraction.
- **Complement**. If A is a set, then $\overline{A}$ stands for the set of all elements that are not in A. One needs to be careful about context with this notation; that is, there needs to be a clearly articulated "universe" of elements. For example, if the context is the integers and X is the set of odd integers, then $\overline{X}$ is the set of even integers. Generally, it is better to use set difference, like this: $\mathbb{Z} - X$.
- **Symmetric difference**. If A and B are sets, then $A \triangle B$ is the set of all elements that are in A or B *but not both*.

$$\{1, 2, 3\} \triangle \{2, 4, 6\} = \{1, 3, 4, 6\}.$$

Observe that $A \triangle B = (A \cup B) - (A \cap B) = (A - B) \cup (B - A)$.
- **Cardinality**. If A is a set, then $|A|$ denotes the number of elements in A. This is also sometimes written #A. For example, if $A = \{1, 2, 4\}$, then $|A| = 3$.
- **Cartesian product**. Given sets A and B, their *Cartesian product* is the set

$$A \times B = \{(a, b) : a \in A, b \in B\}.$$

For example,

$$\{1, 2, 3\} \times \{3, 4\} = \{(1, 3), (1, 4), (2, 3), (2, 4), (3, 3), (3, 4)\}.$$

The notation A^2 stands for $A \times A$; that is, the set of all ordered pairs (x, y) with $x, y \in A$.

More generally, A^n (where n is a positive integer) is the set of all n-long lists of elements of A. See the *Lists* section on the next page.
- **Power set**. If A is a set, then the *power set* of A is the set of all subsets of A. This is typically denoted 2^A or $\mathcal{P}(A)$. For example, if $A = \{1, 2, 3\}$ then

$$\mathcal{P}(A) = 2^A = \Big\{\{1, 2, 3\}, \{1, 2\}, \{1, 3\}, \{2, 3\}, \{1\}, \{2\}, \{3\}, \emptyset\Big\}.$$

- **Set exponentiation**. If A and B are sets, then B^A stands for the set of all functions from A to B; that is,

$$B^A = \{f \mid f : A \to B\}.$$

See page 25 for an explanation of the notation $f : A \to B$.

The rationale for this notation is as follows. If A and B are finite sets with $|A| = a$ and $|B| = b$, then there are b^a functions from A to B. In symbols: $\left|B^A\right| = |B|^{|A|}$.

- **Maximum and minimum**. If A is a set of real numbers, then $\max A$ stands for the largest element in A and $\min A$ stands for the smallest element in A. The wedge and vee symbols are also used in this context:

$$x \vee y = \max\{x, y\} \qquad \text{and} \qquad x \wedge y = \min\{x, y\}.$$

Not all sets of numbers necessarily contain a maximum or minimum element. A related pair of notions are the *supremum* and *infimum* denoted $\sup A$ and $\inf A$. The supremum is also called the *least upper bound* and the infimum is called the *greatest lower bound*; these are sometimes abbreviated lub and glb.

2. Lists

In mathematics, a list is an ordered collection of objects in which repetition is permitted. A list is usually enclosed in round parentheses (although sometimes square brackets are used). For example, $(1, 2, 2, 3)$ is a list. The lists $(1, 2, 2, 3)$, $(1, 2, 3)$, and $(2, 1, 2, 3)$ are all different as order matters and elements may be repeated.

A list of n elements is sometimes called an *n-tuple*.

When a list is named by a letter (e.g., a), the elements of that list are typically named a_1, a_2, etc.

3. Big sums, products, and so on

The symbols $\sum$ and $\prod$ are used to represent the sum and product of a collection of numbers. A typical use of the sum notation has the following form:

$$\sum_{j=\text{start}}^{\text{stop}} \text{expression involving } j.$$

The letter j is a dummy variable. For example,

$$\sum_{j=1}^{5} x^j = x^1 + x^2 + x^3 + x^4 + x^5.$$

If the upper index in the sum is ∞, this means that the sum goes on without end:

$$\sum_{n=0}^{\infty} \frac{1}{2^n} = 1 + \frac{1}{2} + \frac{1}{4} + \frac{1}{8} + \frac{1}{16} + \cdots.$$

The $\prod$ notation is exactly analogous to the $\sum$ notation, but signifies multiplication. For example,

$$\prod_{j=0}^{5} (2j + 1) = 1 \times 3 \times 5 \times 7 \times 9 \times 11.$$

In these examples the sum/product index (dummy variable) traverses a contiguous stretch of integers. However, sometimes we may wish to sum over other kinds of indices.

For example, if A is the set $\{1, 5, 6, 22\}$ then

$$\sum_{t \in A} t^2 = 1^2 + 5^2 + 6^2 + 22^2.$$

Alternatively, a brief description of the desired indices may be written below the sum or product symbol. For example:

$$\prod_{p \text{ prime}} \left(1 - \frac{1}{p}\right) = \left(1 - \frac{1}{2}\right) \times \left(1 - \frac{1}{3}\right) \times \left(1 - \frac{1}{5}\right) \times \left(1 - \frac{1}{7}\right) \times \left(1 - \frac{1}{11}\right) \times \cdots.$$

Sometimes a summation expression is presented without any indication as to which variable is the dummy variable and without specifying upper/lower limits for that index. In such situations, it is often the case that a repeated variable is the summation index. For example, if you see

$$\sum \binom{n}{k} x^k$$

then it is likely that k is the summation index. One then has to infer the proper range for k. In this case, since $\binom{n}{k}$ is defined only for $k \geq 0$ and is zero for $k > n$, it's a safe bet that k ranges from 0 to n. In general, if no upper or lower bounds are given for the a index, then the sum is over all allowable values for that index. For example, if A is an $n \times m$-matrix and B is an $m \times p$-matrix, then

$$\sum a_{i,j} b_{j,k}$$

likely means that the sum is for $j = 1, 2, 3, \ldots, m$.

In some contexts, it is convenient write sums without the $\sum$ symbol altogether by using *Einstein notation*. In this notation, the $\sum$ symbol is omitted. Any time an index (typically a subscript) is repeated, one sums over the range of that index. For example, the term $a_{i,j} b_j$ means

$$\sum_j a_{i,j} b_j$$

where the range of j is (we hope) clear from context.

Following this convention, if A is a square matrix, then $a_{i,i}$ would be the trace of A. For appropriately shaped matrices A and B, $a_{i,j} b_{j,k}$ would denote the i, k-entry of AB.

Analogues of the $\sum$ and $\prod$ notation are used for other operations. For example if $A_1, A_2, A_3, \ldots$ are sets, their union may be written like this:

$$\bigcup_{i=1}^{\infty} A_i$$

which means, of course, $A_1 \cup A_2 \cup A_3 \cup \cdots$. Likewise $\bigcap_{i=1}^{\infty} A_i$ is their intersection.

In general, most operation symbols may be written large with a dummy variable index. For example, if $p_1, p_2, \ldots, p_n$ are Boolean (true/false) values (see Chapter 3), then

$$\bigwedge_{i=1}^{n} p_i$$

means $p_1 \wedge p_2 \wedge \cdots \wedge p_n$.

Logic

1. Boolean operations and proof symbols

The mathematical words *and*, *or*, *not*, *implies*, and so on are often abbreviated with special symbols.

- **And**. $p \wedge q$ is the customary notation for the sentence "p and q." Sometimes an ampersand & is used.
- **Or**. $p \vee q$ is the customary notation for the sentence "p or q." Sometimes a vertical stroke | is used.
- **Not**. The sentence "not p" is denoted $\sim p$ or $\neg p$.

    Note that many math relation symbols may be "slashed" to mean that the relation does not hold. For example $\neq$ (not equal), $\notin$ (not an element of), $\nsubseteq$ (not a subset of), and so on.
- **Exclusive or**. $p \veebar q$ means "p or q, but not (p and q)." Sometimes a circled plus sign $\oplus$ is used. The word XOR is often used as an abbreviation for *exclusive or* and sometimes is used as a notation for this operation: p xor q.
- **Nand**. $p \barwedge q$ means $\neg(p \vee q)$. The term *nand* is a contraction of *not and*.
- **Implies**. The notation $a \Rightarrow b$ means "If a then b" or "a implies b."
- **Implied by**. The notation $a \Leftarrow b$ means "a is implied by b" or "If b then a."
- **If and only if**. The notation $a \Longleftrightarrow b$ means "a if and only if b." In other words "If a then b" and "If b then a" are both true statements. The words *if and only if* are often abbreviated iff.
- **Therefore**. A small triangle of three dots $\therefore$ is an abbreviation for the word *therefore*.
- **Because**. An inverted triangle of three dots $\because$ is an abbreviation for the word *because*.
- **Contradiction**. The notation $\Rightarrow\Leftarrow$ indicates a contradiction has been reached in a proof. It is an abbreviation for: "We have reached a contradiction. Therefore the supposition is false."
- **End of proof**. A square $\square$ (or a filled square $\blacksquare$) is often used to show the end of a mathematical proof. The letters QED are also used; they

stand for *Quod Erat Demonstrandum* [thus it has been demonstrated]. In class, I use this symbol: ☺.

2. Quantifiers

The mathematical symbols ∀ and ∃ are abbreviations that loosely translate to *always* and *sometimes*.

The symbol ∀ is called the *universal quantifier* and means *for all*. For example,

$$\forall x \in \mathbb{R}, \ x^2 \geq 0.$$

This translates to: Every real number x has the property that x^2 is nonnegative. Or, more colloquially, the square of a real number must be zero or larger.

The symbol ∃ is called the *existential quantifier* and means *there exists*. For example,

$$\exists x \in \mathbb{R}, \ x^2 = 2.$$

This sentence asserts that there is a real number whose square equals 2. (And, of course, $\sqrt{2}$ is such a number.)

To make this notation a bit more pronounceable, some people insert the words "such that" (typically abbreviated s.t. or notated ∋) into quantifier notation:

$$\exists x \in \mathbb{R} \text{ s.t. } x^2 = 2$$

reads "there is a real number x such that $x^2 = 2$."

When an ∃ is followed by an exclamation point it means there is a *unique* object that satisfies the property. For example, $\exists! x \in \mathbb{R}, \ x^2 = 0$. This means there is a real number whose square is zero, and there is only one such number.

Quantifiers may be combined to form more complex statements. For example, the property that addition is commutative can be written like this:

$$\forall x \in \mathbb{R}, \ \forall y \in \mathbb{R}, \ x + y = y + x.$$

However, when quantifiers alternate the meaning becomes more difficult to parse. For example,

$$\forall x \in \mathbb{R}, \exists y \in \mathbb{R}, \ x + y = 0$$

asserts that whenever we pick a number x, we can find a number y so that $x + y = 0$. Of course, we should pick $y = -x$. But

$$\exists y \in \mathbb{R}, \forall x \in \mathbb{R}, \ x + y = 0$$

makes the false assertion that there's a "magic" number y with the property that no matter what number x you add to it, the result must be zero.

CHAPTER 4

Numbers

1. Real numbers

Writing numbers. The set of all real numbers is denoted $\mathbb{R}$. Real numbers are typically written in decimal notation starting with a sign (optional if positive, mandatory if negative), a finite list of digits, a decimal point, and then either finitely many or infinitely many more digits.

An infinite repeating block of decimals is often denoted with an overline:

$$45.07123123123123\ldots = 45.07\overline{123}$$

When a decimal number is between 0 and 1, it is preferable to include a leading zero digit. Thus 0.123 is preferred to .123 as the leading zero alerts the reader to the otherwise easily missed decimal point.

For numbers with more than three digits to the left of the decimal point, commas are used to improve readability: 1,332,443. However, for numbers between 1000 and 9999 the comma is often omitted. In some cases, a small space is used instead of a comma: 1 332 443.

The use of a period as the decimal point is not universal. In many countries a comma is used instead (e.g., $\pi \approx 3,14159$) and the period is used to separate groups of three digits (e.g., 1.332.443).

The per cent symbol means "divided by 100". The following numbers are exactly the same; they are simply written differently:

$$\frac{1}{4} \qquad 0.25 \qquad 25\%$$

Scientific notation is often used to express real numbers, especially if they are very large or very small. For example:

$$1.23 \times 10^7 = 12{,}300{,}000 \quad \text{and} \quad 4.56 \times 10^{-4} = 0.000456$$

The number before the power of 10 is usually at least 1 and less than 10. Computer programs may output numbers in scientific notation with an E (or e) to mark the power of ten, like this:

$$1.23 \times 10^7 \qquad \texttt{1.23E07}$$
$$4.56 \times 10^{-4} \qquad \texttt{4.56E-04}$$

Engineering notation is a variant form of scientific notation in which the exponent on 10 must be a multiple of three, like this:

$$\text{Scientific notation:} \qquad 6.022 \times 10^{23}$$
$$\text{Engineering notation:} \quad 602.2 \times 10^{21}$$

For engineering notation, the factor in front of the multiple of 10 should be at least 1 and less than 1000.

Sometimes it is useful to express numbers in bases other than ten. There are a few ways in which this is indicated:

- The base of the number is written as a word subscript like this: 14_{FIVE} or 0.202020_{THREE}.
- The base of the number is written as a numerical subscript, like this: 1011_2.
- In computer science, base-16 (hexadecimal) integers are written with a 0X or 0x prefix. For example: 0X1A2B93 or 0x1a2b93. In this notation, the letters A through F stand for digits whose values are 10 through 15, respectively.
- In computer science, base-8 (octal) integers are written by simply beginning the number with a zero. For example 0177.

Some real numbers are expressed using continued fractions. For example:

$$1 + \cfrac{1}{2 + \cfrac{1}{3 + \cfrac{1}{4 + \cfrac{1}{\ddots}}}}$$

This may be written compactly like this:

$$1 + \frac{1}{2+}\ \frac{1}{3+}\ \frac{1}{4+}\ \cdots\ .$$

In these examples, the numerators are all 1s, but this is not necessary. However, in that case, continued fractions may also be expressed this way: $[1; 2, 3, 4, \ldots]$.

Further number notation. The absolute value of a real number x is denoted $|x|$.

The plus-minus notation $\pm$ is used to indicate two different values. Thus, ± 2 stands for 2 and -2. It is a useful shorthand to encapsulate two different values in a single expression. For example, the solutions to the equation $x^2 - 2x - 2 = 0$ are $1 \pm \sqrt{3}$; this means that both $1 + \sqrt{3}$ and $1 - \sqrt{3}$ are solutions.

Using the symbol $\pm$ twice in a single expression can be ambiguous. Consider $\pm 1 \pm \sqrt{5}$. This might either mean the two values $1 + \sqrt{5}$ and $-1 - \sqrt{5}$, or it might also include the additional values $1 - \sqrt{5}$ and $-1 + \sqrt{5}$. The meaning needs to be derived from context.

The minus-plus symbol $\mp$ means the same thing as $\pm$. It is typically used in conjunction with $\pm$ and has the opposite sign. Thus $\pm 3 \mp \sqrt{7}$ means the two values $3 - \sqrt{7}$ and $-3 + \sqrt{7}$.

The equal sign $=$ has two meanings. One asserts that two values are equal as in $3 + 4 = 7$. The other meaning occurs when defining values as in "Let $x = 1 + \sqrt{5}$." Some people use the notation $:=$ when defining a value. The symbols $\triangleq$ and $\stackrel{\text{def}}{=}$ are also used as defining equal signs.

A triple line equal sign $\equiv$ is used to mean *identically equal to*. For example, an equation may be written as $x^2 - 2 = 0$ which is true for *some* values of x. However, when we write $\sin^2 x + \cos^2 x \equiv 1$, we assert the equation is true for *all* values of x.

2. Subsets of the reals

The integers are the real numbers that can be expressed without any digits after the decimal point (in which case writing the decimal point is optional). The set of integers is denoted $\mathbb{Z}$:

$$\mathbb{Z} = \{\ldots, -3, -2, -1, 0, 1, 2, 3, \ldots\}$$

There is no standard definition for the term *natural number*. We prefer the definition that a *natural number* is a nonnegative integer. Thus $\mathbb{N} = \{0, 1, 2, 3, \ldots\}$. However, some mathematicians prefer not to include 0 in this set.[1]

The *rational numbers* are those real numbers that can be expressed as the ratio of integers a/b where b is nonzero. The set of rational numbers is denoted $\mathbb{Q}$. This may be written

$$\mathbb{Q} = \{x \in \mathbb{R} : x = a/b \text{ where } a, b \in \mathbb{Z} \text{ and } b \neq 0\}.$$

These sets of numbers are nested as follows:

$$\mathbb{R} \supset \mathbb{Q} \supset \mathbb{Z} \supset \mathbb{N}.$$

Appending a star superscript to these may have various meanings:

- $\mathbb{R}^*$, $\mathbb{C}^*$, and so on denote the nonzero elements of the set. [Note: $\mathbb{C}$ stands for the set of complex numbers; see §4.4.]
- $\mathbb{R}^*$, $\mathbb{C}^*$, and so on denote the invertible elements of the set. In the case of the fields, $\mathbb{R}$, $\mathbb{C}$, $\mathbb{Q}$, this is the same as the nonzero elements. In the case of the integers, $\mathbb{Z}^*$ would mean $\{-1, 1\}$. For this meaning, we prefer a $\times$ superscript: $\mathbb{Z}^{\times}$.
- $\mathbb{C}^*$ is sometimes used to denote $\mathbb{C} \cup \{\infty\}$ (though we recommend $\hat{\mathbb{C}}$ or $\overline{\mathbb{C}}$ for this). Similarly, $\mathbb{R}^*$ is sometimes used to denote $\mathbb{R} \cup \{-\infty, +\infty\}$ (but we prefer $\overline{\mathbb{R}}$).

---

[1]The 1993 ANSI/IEEE standard [8] excludes 0 from $\mathbb{N}$ whereas the ISO standard [9] includes 0. The ISO recommends $\mathbb{N}^*$ for the set of positive integers.

- $\mathbb{Z}^*$ is used by [13] to denote the nonnegative integers, $\{0, 1, 2, \ldots\}$ (but we prefer $\mathbb{N}$).
- Finally, $^*\mathbb{R}$ denotes the nonstandard reals (see page 19).

Appending a + superscript or subscript to $\mathbb{R}$ generally denotes the positive reals: $\mathbb{R}^+$ or $\mathbb{R}_+$. However, it is sometimes convenient to include 0; in that way, $\mathbb{R}^n_+$ would denote the nonnegative orthant. Whether or not to include 0 in this set is a matter of convenience to the matter being discussed. If in doubt, ask. Likewise $\mathbb{Q}^+$ and $\mathbb{Z}^+$ denote the positive rationals and integers respectively.

Intervals of real numbers are denoted with open/close parentheses and brackets. A parenthesis indicates that the endpoint is not included whereas a bracket indicates that the endpoint is included. Some examples:

$$[1, 2] = \{x \in \mathbb{R} : 1 \le x \le 2\}$$
$$[1, 2) = \{x \in \mathbb{R} : 1 \le x < 2\}$$
$$(1, 2] = \{x \in \mathbb{R} : 1 < x \le 2\}$$
$$(1, 2) = \{x \in \mathbb{R} : 1 < x < 2\}.$$

Some people use brackets facing the wrong direction to indicate the non-inclusion of an interval's endpoint. For example $[a, b[$ means the same thing as $[a, b)$, namely the set $\{x : a \le x < b\}$. Likewise, $]a, b]$ is the interval $(a, b]$ and $]a, b[$ is the open interval (a, b).

The symbols $-\infty$ and ∞ may be used for the left and right ends of an interval:

$$[1, \infty) = \{x \in \mathbb{R} : x \ge 1\} \quad \text{and} \quad (-\infty, 2) = \{x \in \mathbb{R} : x < 2\}.$$

Although one could write $(-\infty, \infty)$, this would indicate all real numbers and it's clearer simply to write $\mathbb{R}$.

3. "Famous" real numbers

Some real numbers have their own special notation. Here are a few that you may encounter.

- e, the base of the natural logarithms. Its approximate value is 2.71828.

- π, the ratio of a circle's circumference to its diameter. Its approximate value is 3.14159.
- γ, the Euler-Mascheroni constant:

$$\gamma = \lim_{n \to \infty} \left[\sum_{k=1}^{n} \frac{1}{k} - \ln n \right].$$

Its approximate value is 0.5772.
- ϕ, the golden ratio: $\phi = (1 + \sqrt{5})/2$. Its approximate value is 1.618.

Note that various scientific disciplines reserve certain letters for physical constants (such as c for the speed of light).

4. Complex numbers

The complex numbers are created by appending a new object, i, to the real numbers and following the natural consequence of algebraic steps. Here, i stands for a number with the property $i^2 = -1$. The result is a collection of numbers of the form $a + bi$ where $a, b \in \mathbb{R}$. The set of complex numbers is denoted $\mathbb{C}$. Some people prefer to write the imaginary unit before its coefficient, like this: $a + ib$.

[Special note: Electrical engineers use i to represent current, and so they use the letter j to represent $\sqrt{-1}$. For them, complex numbers are written as $a + bj$.]

Just as real numbers are visualized as a (number) line, the complex numbers are visualized as a (complex) plane in which the number $a + bi$ corresponds to the point with coordinates (a, b).

The *absolute value* of $z = a+bi$ is denoted $|z| = |a+bi|$ and equals $\sqrt{a^2 + b^2}$. It is the distance from the point (a, b) to the origin. This is also called the complex number's *magnitude*.

The *(complex) conjugate* of the complex number $z = a + bi$ is denoted with an overline: $\bar{z} = \overline{a + bi} = a - bi$. The conjugate of z is also denoted z^*.

A fancy R and a fancy I are used to denote the real and imaginary parts of a complex number. If $z = a + bi$, then

$$\mathfrak{R}z = a \qquad \text{and} \qquad \mathfrak{I}z = b.$$

The abbreviations Re and Im are also used:

$$\operatorname{Re} z = a \qquad \text{and} \qquad \operatorname{Im} z = b.$$

Points in the plane may also be expressed in polar coordinates (r, θ) where r is the radius/distance of the point from the origin and θ is the counterclockwise angle of the point from positive x-axis. If a point has rectangular coordinates (a, b) and polar coordinates (r, θ), then these quantities are related by the equations

$$a = r\cos\theta \qquad \text{and} \qquad b = r\sin\theta.$$

This idea extends to complex numbers in that every complex number may be expressed as $re^{i\theta}$ where $r, \theta \in \mathbb{R}$ because

$$re^{i\theta} = r[\cos\theta + i\sin\theta] = (r\cos\theta) + (r\sin\theta)i = a + bi.$$

Thus $r = |a + bi|$. The angle θ is called the *argument* or the *phase angle* of $a+bi$; this is denoted $\arg(a+bi)$. Typically one defines $\arg z$ to lie in the interval $[0, 2\pi)$ or in the interval $(-\pi, \pi]$. See Figure 4.1 on the following page.

The value $e^{i\theta}$ is sometimes written $\operatorname{cis}\theta$. The notation cis is an abbreviation for cos plus i sin:

$$\operatorname{cis}\theta = \cos\theta + i\sin\theta.$$

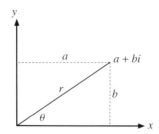

FIGURE 4.1. Rectangular and polar representation of a complex number $a + bi = re^{i\theta}$.

5. Basic operations

For two numbers a and b their sum is written $a + b$ and their difference is $a - b$.

Multiplication is expressed in several different ways. If the numbers are expressed as letters, then ab is the most commonly used notation. For the product of a specific number and a letter, the number is usually written first, e.g., $3x$. The product of two specific numbers is written with either $\cdot$ or $\times$ to show the operation: $5 \cdot 7$ or 8×11. Because the $\cdot$ and $\times$ symbols are not standard keyboard characters, some people (and most computer languages) use an asterisk to stand for multiplication: $3 * 5$.

Division is expressed in one of the following ways: $a \div b$, a/b, or $\frac{a}{b}$. One should choose a notation that maximizes readability.

Exponentiation is expressed using superscripts: a^b. However, a wedge $\wedge$ or double asterisks $**$ are sometimes used.

See the discussion of $\sum$ and $\prod$ notation on page 8.

6. Other number systems

There are other number systems used by mathematicians that are useful in engineering and science applications. We list here some of the more common ones.

- **Modular numbers**. For an integer $n \geq 2$, we write $\mathbb{Z}_n$ to stand for the set $\{0, 1, \ldots, n - 1\}$ with operations mod n. The notations $\mathbb{Z}/n\mathbb{Z}$ and $\mathbb{Z}/(n)$ are also used.
- **Finite fields**. A finite field[2] with n elements may be written either GF(n) or $\mathbb{F}_n$. It is known that the size of a finite field must be a power of a prime, so it is likely that this notation will be seen like this: GF(p^a) or $\mathbb{F}_{p^a}$.

---

[2]Roughly speaking, a *field* is a number system in which the four basic operations obey the same algebraic properties as they do for the real numbers. The most common fields encountered are $\mathbb{R}$, $\mathbb{Q}$, and $\mathbb{C}$. Consult an abstract algebra text for the full story.

- **Gaussian integers**. The *Gaussian integers* are numbers of the form $a + bi$ where a and b are integers. The set of Gaussian integers is denoted $\mathbb{Z}[i]$.
- **Quaternions**. Hamilton's *quaternions* are numbers of the form $a + bi + cj + dk$ where $a, b, c, d \in \mathbb{R}$ and i, j, k have the following properties:

$$i^2 = -1 \qquad\qquad ij = k \qquad\qquad ik = -j$$
$$j^2 = -1 \qquad\qquad ji = -k \qquad\qquad jk = i$$
$$k^2 = -1 \qquad\qquad ki = j \qquad\qquad kj = -i.$$

The set of all quaternions is denoted $\mathbb{H}$. Note that multiplication in $\mathbb{H}$ is not commutative.

- **Extensions**. Above we presented the notation $\mathbb{Z}[i]$ for the Gaussian integers. This notation means we take the integers $\mathbb{Z}$ together with the imaginary number i and build all possible numbers using the operations addition, subtraction, and multiplication.

  This notation generalizes to any ring[3] and any auxiliary element α we wish to append to the ring. The notation $R[\alpha]$ denotes the set of all objects we can form by repeatedly using addition, subtraction, and multiplication with the elements of R and α. For example, $\mathbb{Z}[\sqrt{-2}]$. See also the notation $R[x]$ in §6.4 on polynomials.

  For a field F, the notation $F(\alpha)$ is the extension of F created by including the element α with repeated use of all four basic operations. Thus $\mathbb{Q}(\pi)$ includes numbers such as $\frac{2+5\pi}{1/2-\pi^2}$.

  See page 35 for the notation $R[\![x]\!]$, the set of formal power series with coefficients from R in the variable x.

7. To infinity and beyond

It is often useful to append the concept of infinity to the real or complex number systems.

In the realm of real numbers, $\overline{\mathbb{R}}$ denotes the set of *extended real numbers* which includes the additional values $+\infty$ and $-\infty$.

For complex numbers, $\hat{\mathbb{C}}$ [or $\overline{\mathbb{C}}$] denotes the *extended complex numbers* (also called the *Riemann sphere*) which includes the single additional value ∞. This system is sometimes expressed simply as $\mathbb{C} \cup \{\infty\}$. Some authors write $\mathbb{C}^*$, but this is also (and preferably) used to denote the set $\mathbb{C} - \{0\}$.

An exotic extensions of the real numbers is $^*\mathbb{R}$, the *nonstandard reals* which includes infinitesimals and hyperintegers.

---

[3]Roughly speaking, a *ring* is a number system in which the operations addition, subtraction, and multiplication obey the same properties they do for ordinary numbers. However, division is not necessarily defined. The set of integers $\mathbb{Z}$ is a good example of a ring.

As discussed on page 7, given a set A, the notation $|A|$ gives the cardinality (size) of the set A. This is an integer for finite sets. There is, however, special notation for infinite sets.

The symbol $\aleph_0$ denotes the cardinality of the integers: $\aleph_0 = |\mathbb{Z}|$. It is the smallest transfinite cardinal number. The symbol $\aleph$ is the Hebrew letter *aleph* and the notation $\aleph_0$ is usually spoken "aleph null" or "aleph naught". Sets with cardinality $\aleph_0$ are called *countable*.

The cardinality of the real numbers, $|\mathbb{R}|$, is denoted c. The quantity c is also called the *cardinality of the continuum*.

CHAPTER 5

Geometry

1. Fundamentals

The most basic geometric object is a point and points are often denoted with upper case letters, A, B, C, etc. However, points in the plane can be represented as 2-tuples (x, y) or as vectors $\left[\begin{smallmatrix} x \\ y \end{smallmatrix}\right]$; as such, they may be named like vectors with lower case bold letters, $\mathbf{x}$, $\mathbf{y}$, and so on. The entire Euclidean plane may be written as $\mathbb{R}^2$, but for those who prefer to think of the plane in a coordinate-free, synthetic way, we may write E^2 or $\mathbb{E}^2$. More generally, n-dimensional Euclidean space is $\mathbb{R}^n$, E^n, or $\mathbb{E}^n$.

Two distinct points A and B determine a unique line which is typically denoted $\overleftrightarrow{AB}$. The ray emanating from A and including B is denoted $\overrightarrow{AB}$. The line segment joining A and B is denoted $\overline{AB}$, however, for simplicity's sake, some people write AB for the line segment. Line segments have length which can be denoted in various ways including the same notation as for the segment itself (AB or $\overline{AB}$) or with absolute value bars, $|AB|$ or $|\overline{AB}|$.

An *angle* is the union of two rays emanating from the same point. If the rays are $\overrightarrow{AB}$ and $\overrightarrow{AC}$, the angle can be denoted $\angle BAC$ or, if there is only one angle with vertex A under consideration, $\angle A$. Sometimes angles in diagrams are marked with numbers or Greek letters, so one may see $\angle 1$ or $\angle \alpha$. The measure of an angle is written $m\angle BAC$, but some people omit the m and use the same notation for an angle and its measure.

A triangle is the union of three line segments determined by three points and is denoted $\triangle ABC$.

In a triangle, the angles are named by their vertex points ($\angle A$, $\angle B$, $\angle C$) and the lengths of the sides are named by lower case letters corresponding to the opposite angle. That is, a is the length of the segment opposite $\angle A$ (the length of the segment $\overline{BC}$). See Figure 5.1. A classic example of this is the law of cosines:

$$c^2 = a^2 + b^2 - 2ab \cos C.$$

Here c is the length of the segment AB and $\cos C$ is the cosine of the (measure of) $\angle C$.

There are some standard relation symbols for geometric objects. The most basic is equality, denoted with an equals sign $=$, which means the two things are exactly the same. More generally, two geometric figures are *congruent* if there

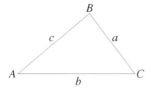

FIGURE 5.1. Standard labeling of a triangle.

is an isometry of space that equates one with the other. Congruence of figures is denoted with the symbol $\cong$, e.g., $\angle A \cong \angle Y$. More general still is similarity of figures which is denoted $\sim$, e.g., $\triangle ABC \sim \triangle DEF$.

Lines, rays, and segments may be *parallel*; this is denoted $\overleftrightarrow{AB} \parallel \overleftrightarrow{XY}$. They may also be *perpendicular*: $\overleftrightarrow{AB} \perp \overleftrightarrow{CD}$.

2. Coordinates

In analytic geometry, points are specified by a list of numbers. The most common system for doing this is known as *Cartesian coordinates* (also called *rectangular coordinates*). Points are specified by d-tuples (i.e., lists of d numbers) either written horizontally (x, y) or vertically $\left[\begin{smallmatrix} x \\ y \end{smallmatrix}\right]$. It is traditional to use the letters x and y for Cartesian coordinates in two dimensions and x, y, and z in three dimensions. In higher dimensions, it is easiest to use subscripts, e.g., (x_1, x_2, x_3, x_4).

The Cartesian convention is not the only system of coordinates. Here are a few others for 2- and 3-dimensional space and their associated notation.

- **Polar coordinates**. In the plane, points may be specified by a distance from a fixed origin and a rotation from a ray emanating from that origin (typically the positive x-axis). It is traditional to use the letters (r, θ) where r is the distance and θ is the rotation. The relation to Cartesian coordinates is
$$x = r \cos \theta \quad \text{and} \quad y = r \sin \theta.$$

- **Spherical coordinates**. *Spherical coordinates*, also known as *spherical polar coordinates*, are a natural extension of polar coordinates to three-dimensional space. Each point is specified by a distance r from the origin, a first rotation clockwise θ from the $+x$-axis in the xy-plane, and a second rotation ϕ out of the xy-plane.

  The point is written (r, θ, ϕ) and θ is called the *azimuth* and ϕ is called the *inclination* or *polar angle*. However, some authors use a different order for the two angles in the triple.

  The conversion from spherical to Cartesian coordinates is given by these formulas:
$$x = r \cos \theta \sin \phi, \quad y = r \sin \theta \sin \phi, \quad \text{and} \quad z = r \cos \phi.$$

- **Cylindrical coordinates**. This is another extension of polar coordinates. Points in space are first located in the x, y-plane by polar coordinates (r, θ) and then a height above that plane by z. Thus the full specification is the triple (r, θ, z).

  The conversion to Cartesian coordinates is

  $$x = r \cos \theta, \quad y = r \sin \theta, \quad \text{and} \quad z = z.$$

3. Differential geometry

Curves. A *curve* is (the image of) a continuous function $\alpha : [a, b] \to \mathbb{R}^k$. This parameterization is *unit speed* if $\|\alpha'(s)\| = 1$ for all $s \in [a, b]$.

Associated with curves are the following notations:

- $\mathbf{T}(s)$ is the *tangent vector* at $\alpha(s)$. For a unit speed curve, we have $\mathbf{T}(s) = \alpha'(s)$.
- $\mathbf{N}(s)$ is the *normal vector* at $\alpha(s)$. For a unit speed curve, we have $\mathbf{N}(s) = T'(s)/\|T'(s)\|$.
- $\mathbf{B}(s)$ is the *binormal vector* at $\alpha(s)$. For a curve in $\mathbb{R}^3$, we have $\mathbf{B}(s) = \mathbf{T}(s) \times \mathbf{N}(s)$.
- $\kappa(s)$ is the *curvature* of the curve at $\alpha(s)$. For a unit speed curve, $\kappa(s) = \|T'(s)\|$.
- $\tau(s)$ is the *torsion* of the space curve at $\alpha(s)$. It is given by the formula $\tau(s) = -\mathbf{B}'(s) \cdot \mathbf{N}(s)$.

Surfaces. Various geometric surfaces have special notation used to describe them.

- **Spheres**. The sphere is S^2. In general S^k is a k-dimensional sphere, so S^1 is a circle.

  A sphere is the boundary of a ball. The notation $B(\mathbf{x}, r)$ is often used for the ball centered at $\mathbf{x}$ with radius r, i.e., $B(\mathbf{x}, r) = \{\mathbf{y} : \|\mathbf{y} - \mathbf{x}\| \le r\}$.
- **Torus**. The following notations are used to denote a torus (the surface of a doughnut): T^2, $S^1 \times S^1$, and $\mathbb{R}^2/\mathbb{Z}^2$.
- **Projective plane**. The (real) projective plane is denote $\mathbb{RP}^2$.
- **Hyperbolic plane**. The hyperbolic plane is denoted $\mathbb{H}^2$.

The *Euler characteristic* of a surface M is denoted $\chi(M)$. If a connected graph is embedded on the surface of M so that every face is homeomorphic to the interior of a disc, then $\chi(M) = v - e + f$ where v is the number of vertices, e is the number of edges, and f is the number of faces.

The *Gaussian curvature* at a point p on a surface is denoted $K(p)$. The Gaussian curvature of a compact, boundaryless surface M is related to the Euler characteristic by the Gauss-Bonnet formula

$$\int_M K \, dA = 2\pi \chi(M)$$

where dA is an element of surface area.

Gaussian curvature is related to the *principal* curvatures of the surface, which are denoted k_1 and k_2, by the equation $K = k_1 k_2$.

CHAPTER 6

Functions

1. Fundamentals

Basic notation. Many of the functions encountered in science and engineering take real numbers as input and return real numbers as output. The notation $f\colon \mathbb{R} \to \mathbb{R}$ means that f is a function, that any real number may be the input to f, and that the output of f lies in the set of real numbers.

More generally, if A and B are any sets, the notation $f\colon A \to B$ means that f is a function that takes as input values from the set A and returns values in the set B.

It is not necessary that every value of B be realized as an output. For example, we may write $\cos\colon \mathbb{R} \to \mathbb{R}$ because the cosine function takes any real number as input and returns real numbers. Of course, the outputs of the cosine function lie in the interval $[-1, 1]$.

The notation $f(x)$ indicates the value returned by the function f when evaluated at the value x. That is, f is the function while $f(x)$ is the value (typically a number) returned by the function.

An alternative notation for function evaluation is sometimes used, especially when the function is specified by a formula. The general form is this:

$$\text{(some expression involving a variable } x)\Big|_{x=a}$$

This means to evaluate the expression by substituting the value a for the variable x. This type of notation is often used with Leibniz notation derivatives (see page 50).

The notation $y = f(x)$ is sometimes abbreviated to $x \mapsto y$; this is pronounced "x maps to y." Note that the function f is absent from the expression $x \mapsto y$ so it is important to use this notation only when the function under consideration is unambiguously known. Alternatively, one may write $x \overset{f}{\mapsto} y$ to mean $f(x) = y$.

A function whose inputs and outputs are themselves functions is often called an *operator* or a *transform*. It is useful to typographically distinguish operators from functions of numbers; for example, one may capitalize the operator's name. The application of an operator to a function might omit the parentheses: Lf denotes the evaluation of the operator L on the function f.

The notation $f(\cdot)$ is sometimes used to emphasize that f is a function with the dot showing that an argument is expected. This notation is handy when referring to a function specified by non-letter symbols. For example, when discussing the absolute value function, one may write $|\cdot|$; this may be preferable to putting in a dummy variable between the bars or leaving a blank space.

Functions may take more than one argument, e.g., $f(x, y)$ is a function of two arguments. In this case the dot notation is handy in expressing which argument to the function is constant and which is variable. For example, $f(\cdot, b)$ expresses a function of one variable: the first argument of f varies over its domain while the second argument is held constant at the value b.

For a function $f : X \to Y$, the *image* of f is denoted $\text{Im}(f)$:

$$\text{Im}(f) = \{y \in Y \mid \exists x \in X, \ f(x) = y\}.$$

More generally, if $A \subseteq X$, then $f(A)$ stands for the set $\{f(a) \mid a \in A\}$. Thus $f(X) = \text{Im}(f)$.

Functions are often specified by a single algebraic formula, but this is not necessary. A function may be defined piecewise like this:

$$f(x) = \begin{cases} e^{-1/x^2} & \text{for } x > 0 \\ 0 & \text{otherwise.} \end{cases}$$

Composition and inversion. If $f : A \to B$ and $g : B \to C$ then $g \circ f$ is a new function, the *composition* of g and f, with $(g \circ f) : A \to C$ defined by

$$(g \circ f)(x) = g[f(x)].$$

The notation $f^{-1}(y)$ stands for the value x such that $f(x) = y$; we call f^{-1} the *inverse* of f.

If there is more than one value x with $f(x) = y$, then f^{-1} is not a function. Some people use the notation $f^{-1}(y)$ to stand for the set of all x values that map to y. However, in some instances one coerces f^{-1} to be a function by restricting the choice of x. For example, the arc sine function is $\sin^{-1}$. There are infinitely many values x such that $\sin x = \frac{1}{2}$, but we restrict the choice of x (in the case of arc sine) to be in the interval $(-\frac{\pi}{2}, \frac{\pi}{2}]$. In this case, arc sine becomes a proper function. Note that $f^{-1}(y)$ is undefined if there is no value x such that $f(x) = y$.

The notation f^{-1} may be applied to a *set* of values. For example, suppose X and Y are sets and $f : X \to Y$. If $B \subseteq Y$, then

$$f^{-1}(B) = \{x \in X \mid f(x) \in B\}.$$

The notation $f^2(x)$ sometimes means $[f(x)]^2$; this notation is used especially with trigonometric functions, e.g., $\sin^2 \theta$. Thus we have an inconsistency because $f^{-1}(x)$ does *not* mean $1/f(x)$. (See the discussion of trigonometric functions later in this chapter.) However, in some contexts $f^2(x)$ means $(f \circ f)(x) = f[f(x)]$ in which case this notation is consistent with the meaning of f^{-1}.

2. Standard functions

Functions of real numbers. There is a vast array of functions defined on the real numbers. Here we present a modest assortment that have their own special notation.

- **Absolute value**. For a real or complex number x, $|x|$ is the absolute value of x. (See also pages 14 and 17.)
- **Square (and other) roots**. For nonnegative real numbers x, the notation $\sqrt{x}$ is unambiguously the nonnegative square root of x. Likewise, for $x \geq 0$ and $n > 0$, $\sqrt[n]{x}$ is the nonnegative n^{th}-root of x. (Note: n need not be an integer.)

  These may be written in exponential form: $\sqrt{x} = x^{1/2}$ and $\sqrt[n]{x} = x^{1/n}$.

  When x is a negative real number the situation becomes more complex (pun intended). In this case $\sqrt{x} = i\sqrt{|x|}$. If n is an odd integer, $\sqrt[n]{x}$ is the unique (necessarily negative) real number y so that $y^n = x$.

  All other cases are ambiguous and handled by special convention.
- **Signum (sign)**. The notation $\operatorname{sgn} x$ is the sign of x; that is,

$$\operatorname{sgn} x = \begin{cases} 1 & \text{if } x > 0, \\ 0 & \text{if } x = 0, \text{ and} \\ -1 & \text{if } x < 0. \end{cases}$$

  It is called the *signum* function. Some people write the full word sign in place of sgn.
- **Exponential function**. $\exp x$ stands for e^x. Sometimes people enclose the argument in curly braces like this: $\exp\{x\}$, but this is not necessary. If the argument is an algebraic expression, parentheses are a good alternative: $\exp\left(x^2 + y^2\right)$.
- **Logarithms**. The notation $\log x$ is somewhat ambiguous. In science and engineering it typically means the base-10 logarithm. In mathematics, it typically means the base-e (natural) logarithm.

  The notation $\ln x$ stands for $\log_e x$ and $\lg x$ stands for $\log_2 x$.

  The function $\log^* x$ occurs in computer science. It stands for the number of times one must apply the log function to a number to reach a result that is less than 1. It is called the *log star* function.
- **Trigonometric functions**. The standard trigonometric functions are sin, cos, tan, sec, csc, and cot.

  In mathematics, the arguments to trigonometric functions are usually expressed in radians; in science and engineering they might be expressed in degrees.

  Greek letters, especially θ, often appear as the argument to trigonometric functions, but this is a custom, not a requirement.

The notation $\sin^2 x$ means $(\sin x)^2$; however, the notation $\sin^{-1} x$ does not mean $1/\sin x$. Rather, $\sin^{-1} x$ is the inverse sine (or arc sine) function. This may also be written $\arcsin x$ (although the $\sin^{-1} x$ notation is more common). Likewise we have arccos, arctan, and so forth.

- **Hyperbolic trigonometric functions**. The standard functions of trigonometry (sine, cosine, tangent, and so on) have hyperbolic "cousins". These are denoted by appending the letter 'h' after their names. For example:

$$\sinh x = \frac{e^x - e^{-x}}{2} \qquad \text{and} \qquad \cosh x = \frac{e^x + e^{-x}}{2}.$$

  Also: $\tanh x = \sinh x / \cosh x$, $\operatorname{sech} x = 1/\cosh x$, $\operatorname{csch} x = 1/\sinh x$, and $\coth x = \cosh x / \sinh x$.
  As with the ordinary trigonometric functions $\sinh^2 x$ means $(\sinh x)^2$ but $\sinh^{-1} x$ is the inverse function of sinh.
- **Sine cardinal**. The function $\operatorname{sinc} x$ is called the *sine cardinal*. It is variously defined as

$$\operatorname{sinc} x = \frac{\sin x}{x} \qquad \text{or} \qquad \operatorname{sinc} x = \frac{\sin \pi x}{\pi x}$$

  for $x \neq 0$ and $\operatorname{sinc} 0 = 1$.
- **Floor, ceiling**. The notation $\lfloor x \rfloor$ stands for the *floor* of x. This is the result of rounding x down to the nearest integer. In older books this is written $[x]$.
  The notation $\lceil x \rceil$ is the *ceiling* of x; it is the result of rounding x up to the nearest integer. In older books this is written $\{x\}$.
  Thus,

$$\lceil e \rceil = \lceil 3 \rceil = \lfloor 3 \rfloor = \lfloor \pi \rfloor = 3.$$

- **Error function**. $\operatorname{erf} x$ is a function that occurs in probability theory. It is defined by

$$\operatorname{erf} x = \frac{2}{\sqrt{\pi}} \int_0^x \exp\left\{-t^2\right\} dt.$$

- **Gamma function**. $\Gamma(x)$ is defined by

$$\Gamma(x) = \int_0^\infty t^{x-1} e^{-t} dt.$$

  It is closely related to the factorial function.
- **Beta function**. $B(a, b)$ is defined by

$$B(a, b) = \int_0^1 t^{a-1} (1 - t)^{b-1} dt$$

  and is equal to $\Gamma(a)\Gamma(b)/\Gamma(a + b)$. It is closely related to binomial coefficients.

- **Bessel functions**. The following notations are used for Bessel functions:
  - J_n: Bessel function of the first kind, order n.
  - Y_n: Bessel function of the second kind, order n.
  - I_n: modified Bessel function of the first kind, order n.
  - K_n: modified Bessel function of the second kind, order n.

  Closely related to these are the Hankel functions of the first and second kinds which are, respectively:

  $$H_n^{(1)}(x) = J_n(x) + iY_n(x) \quad \text{and} \quad H_n^{(2)}(x) = J_n(x) - iY_n(x).$$

- **Hypergeometric functions**. The hypergeometric function $_2F_1$ takes three "parameters" (a, b, and c) and one argument (x):

  $$_2F_1(a, b; c; x) = \sum_{k=0}^{\infty} \frac{a^{(k)} b^{(k)}}{c^{(k)}} \cdot \frac{x^k}{k!}.$$

  For an explanation of the rising factorial notation $a^{(k)}$ see the entry on factorial functions on the next page.

  The notation $_2F_1(a, b; c; x)$ is also more elaborately written like this:

  $$_2F_1 \left(\begin{matrix} a, b \\ c \end{matrix} \middle| x \right)$$

  This notation more clearly illustrates the placement of a and b in the numerator and c in the denominator. The LaTeX code (which relies on the amsmath package) to produce this formula is:

  ```
  {_2}F_1 \left( \genfrac{.}{|}{0pt}{}{a,b}{c} x \right)
  ```

  More generally, if p and q are nonnegative integers, the function $_pF_q$ has p upper indices (a_1 through a_p) and q lower indices (b_1 through b_q) and is notated like this:

  $$_pF_q(a_1, a_2, \ldots, a_p; b_1, b_2, \ldots, b_q; x) = {_pF_q} \left(\begin{matrix} a_1, a_2, \ldots, a_p \\ b_1, b_2, \ldots, b_q \end{matrix} \middle| x \right)$$

  and is given by the formula

  $$\sum_{k=0}^{\infty} \frac{a_1^{(k)} a_2^{(k)} \cdots a_p^{(k)}}{b_1^{(k)} b_2^{(k)} \cdots b_q^{(k)}} \cdot \frac{x^k}{k!}.$$

- **Riemann zeta function**. The function $\zeta(z)$ is known as the *Riemann zeta function*. For an integer $n \geq 2$ we have

  $$\zeta(n) = \sum_{k=0}^{\infty} \frac{1}{k^n}$$

  and is defined for arbitrary complex z by analytic extension.

- **Dirac delta "function"**. The final entry for this section is given with some trepidation because it is *not* a function; however, since the word *function* is traditionally (albeit erroneously[1]) used with its name, we recall here the Dirac δ-function. Dirac's δ can be imagined as a function that takes the value zero for all nonzero arguments, but $\delta(0)$ is infinite. Nevertheless, the integral of δ (the area under its "curve") is 1. The δ function often appears inside an integral; given a function $f \colon \mathbb{R} \to \mathbb{R}$ we have

$$\int_{-\infty}^{\infty} f(x)\delta(x)\,dx = f(0).$$

  The Dirac δ may be thought of as the derivative of the *Heaviside function H* defined by

$$H(x) = \begin{cases} 1 & \text{for } x \geq 0 \\ 0 & \text{for } x < 0. \end{cases}$$

Notice that

$$H(x) = \int_{-\infty}^{x} \delta(t)\,dt$$

and in that sense the derivative of H is δ.

  In case one wants a spike at a value other than 0 (say at c) one can use $\delta(x - c)$ which is sometimes denoted $\delta_c(x)$.

Functions of integers. The following are some functions that are defined only for integer values.

- **Factorial and related functions**. For a nonnegative integer n, the notation $n!$ is pronounced *n-factorial* and is defined by $0! = 1$ and for $n > 0$

$$n! = n(n-1)(n-2)\cdots 2 \cdot 1.$$

  It is closely related to the gamma function: $n! = \Gamma(n+1)$.

  The use of the exclamation point for the factorial function is universally accepted, but it is interesting to note that this was not always the case; see Figure 6.1. For more historical tidbits on notation see [**1**] (and also [**10**]).

$$f(x) = f(a) + \frac{(x-a)}{\underline{1}} f'(a) + \frac{(x-a)^2}{\underline{2}} f''(a) + \frac{(x-a)^3}{\underline{3}} f'''(a) + \cdots$$

FIGURE 6.1. The formula for the Taylor series of a function as it appears in [**4**]. Note the notation for $n!$.

  The notation $n!!$ is the double factorial; here the factors descend by two. For example, $7!! = 7 \times 5 \times 3 \times 1$.

---

[1]Dirac's δ is more properly called a *generalized function* or a *distribution*.

For a real number x and a nonnegative integer k, the notations $x^{(k)}$ and $x_{(k)}$ are known respectively as the *rising* and *falling* factorial functions and are given by the formulas:

$$x^{(k)} = x(x+1)\cdots(x+k-1) = \prod_{j=0}^{k-1}(x+j) \quad \text{and}$$

$$x_{(k)} = x(x-1)\cdots(x-k+1) = \prod_{j=0}^{k-1}(x-j).$$

Note that $x^{(0)} = x_{(0)} = 1$.

Alternative (and especially clear) notations for rising/falling factorials are these:

$$x^{\overline{k}} = x^{(k)} \qquad \text{and} \qquad x^{\underline{k}} = x_{(k)}.$$

In addition, the notations $P(n,k)$, $P_{n,k}$, $P_{k,n}$, and $_nP_k$ are also used for falling factorial $n_{(k)}$, but these are mostly seen in older works. Here the letter P stands for *permutation* and $P(n,k)$ is understood to mean "the number of permutations of n things taken k at a time." We discourage the use of P for this purpose.

Finally, and unfortunately, the notation $(x)_k$ (known as the Pochhammer symbol) is also used for the *rising* factorial function.

• **Binomial coefficient**. The binomial coefficient $\binom{n}{k}$ is defined for nonnegative integers n and k to be the number of k-element subsets of an n-element set. For $0 \le k \le n$ we have

$$\binom{n}{k} = \frac{n!}{k!(n-k)!}$$

and for $k > n$ we have $\binom{n}{k} = 0$.

The letter C is also used for the binomial coefficient in these various forms: $C(n,k)$, $C_{n,k}$, $C_{k,n}$, and $_nC_k$. In this context C stands for *combinations* and $C(n,k)$ stands for "the number of combinations of n things taken k at a time." We grudgingly acknowledge the utility of $C(n,k)$ (especially for communicating mathematics in an email) but strongly encourage the exclusive use of $\binom{n}{k}$.

There are some additional notations closely related to binomial coefficients.

– **Multinomial coefficient**. For nonnegative integers n, $k_1, k_2, \ldots, k_t$ with $k_1 + k_2 + \cdots + k_t = n$ we have

$$\binom{n}{k_1 \; k_2 \; \cdots \; k_t} = \frac{n!}{k_1! k_2! \cdots k_t!}.$$

– **Multichoose**. The notation $\left(\!\binom{n}{k}\!\right)$ stands for the number of k-element multisets that can be formed with elements drawn from an n-element set. The multichoose notation is related to the ordinary binomial coefficient by the following formula:

$$\left(\!\!\binom{n}{k}\!\!\right) = \binom{n + k - 1}{k}.$$

– q-**binomial coefficient**. For nonnegative integers n, k and a variable q the notation $\binom{n}{k}_q$ is the q-*binomial coefficient* defined by

$$\binom{n}{k}_q = \begin{cases} \frac{(1-q^n)(1-q^{n-1})\cdots(1-q^{n-k+1})}{(1-q)(1-q^2)\cdots(1-q^k)} & \text{if } k \leq n \text{ and} \\ 0 & \text{otherwise.} \end{cases}$$

• **Stirling numbers**. Stirling numbers come in two varieties. For nonnegative integers n, k we have:
  – Stirling numbers of the first kind are denoted $s(n, k)$. It is the coefficient of x^k in the polynomial $(x)_n = x(x - 1) \cdots (x - n + 1)$. We use $c(n, k)$ to denote the *unsigned* Stirling numbers of the first kind. It equals $|s(n, k)|$ and is the number of permutations in S_n (the set of all permutations of $\{1, 2, \dots, n\}$—see page 43) that have exactly k cycles. It is also denoted $\left[{n \atop k}\right]$.
  – Stirling numbers of the second kind are denoted $S(n, k)$. They represent the number of partitions of an n-element set into exactly k parts. An alternative notation is $\left\{{n \atop k}\right\}$.
• **Divisibility**. For integers a and $b \neq 0$, the expression $b|a$ means a is divisible by b, i.e., there is an integer q such that $a = bq$. The statement $b|a$ is read "b divides a."
• **Mod**. For integers a, b with $b > 0$ the expression $a \bmod b$ is the remainder when a is divided by b. Specifically, dividing a by b yields integers q (quotient) and r (remainder) such that

$$a = qb + r \quad \text{and} \quad 0 \leq r < b.$$

The value of $a \bmod b$ is the remainder r. For example $13 \bmod 10 = 3$ and $-10 \bmod 9 = 8$ (because $-10 = -2 \times 9 + 8$).

In addition to the function mod, there is also a relation on integers called *congruence* that employs the term mod. For a positive integer n and any integers a and b we write

$$a \equiv b \pmod{n} \qquad \text{or more simply} \qquad a \equiv b \ (n)$$

to mean $a - b$ is divisible by n (i.e., $n|a - b$). When the modulus, n, is understood the $(\bmod\ n)$ or (n) portion may be omitted.
• **GCD and LCM**. For integers n and m, the notation $\gcd(n, m)$ is the greatest common divisor of n and m. In number theory, this is abbreviated to (n, m).

The notation $\text{lcm}(n, m)$ is the least (positive) common multiple of n and m.

- **Euler's totient**. For a positive integer n, $\varphi(n)$ is the number of integers in $[n] = \{1, 2, \ldots, n\}$ that are relatively prime to n. This function is known as Euler's *totient* and the symbol φ is a stylized Greek phi, ϕ. Some people use the ordinary ϕ for this function.
- **Legendre/Jacobi symbol**. If p is an odd prime and a is an integer, the notation $\left(\frac{a}{p}\right)$ is known as the *Legendre symbol* defined by

$$\left(\frac{a}{p}\right) = \begin{cases} 0 & \text{if } a \equiv 0 \pmod{p}, \\ 1 & \text{if } a \text{ is a quadratic residue mod } p \text{ and } a \not\equiv 0, \text{ and} \\ -1 & \text{otherwise.} \end{cases}$$

Some authors write the Legendre symbol horizontally: $(a|p)$.

If the lower entry in $\left(\frac{a}{n}\right)$ is not prime, then this notation is known as the *Jacobi symbol*. If the prime factorization of n is $p_1^{e_1} p_2^{e_2} \cdots p_t^{e_t}$, then

$$\left(\frac{a}{n}\right) = \left(\frac{a}{p_1}\right)^{e_1} \left(\frac{a}{p_2}\right)^{e_2} \cdots \left(\frac{a}{p_t}\right)^{e_t}.$$

Other standard functions.

- **Identity function**. For a set A, the notation Id_A is the *identity function* on A, i.e., $\text{Id}_A : A \to A$ by $\text{Id}_A(a) = a$ for all $a \in A$.
- **Characteristic function**. Given a set A, often a subset of the real numbers, χ_A denotes the *characteristic function* of A given by

$$\chi_A(x) = \begin{cases} 1 & \text{if } x \in A, \text{ and} \\ 0 & \text{otherwise.} \end{cases}$$

Some people write $\mathbf{1}_A$ for the characteristic function of A.

3. Classes of functions

The notation C^k denotes those functions for whose k^{th} derivative is defined and continuous. Thus C^0 stands for the class of continuous functions, C^1 stands for the class of differentiable functions whose derivatives are continuous. Naturally,

$$C^0 \supset C^1 \supset C^2 \supset \cdots .$$

For functions defined only on an interval $[a, b]$, we may write $C^k([a, b])$. The notation $C^k(\mathbb{R})$ is equivalent to C^k.

The class C^∞ contains those functions for which derivatives of all orders exist (and are therefore continuous).

The notation L^p, where p is a positive real number, denotes the class of functions for which

$$\int_{-\infty}^{\infty} |f(x)|^p \, dx < \infty.$$

This notation may be restricted to an interval: $L^p([a, b])$ stands for those functions for which

$$\int_a^b |f(x)|^p \, dx < \infty.$$

Closely related to this is the notation ℓ^p. This stands for the class of sequences $a_0, a_1, a_2, \ldots$ for which

$$\sum_{k=0}^{\infty} |a_k|^p < \infty.$$

4. Polynomials, power series, and rational functions

Polynomials. A *polynomial* is a function of the form

$$p(x) = a_n x^n + a_{n-1} x^{n-1} + \cdots + a_1 x + a_0.$$

Some authors prefer that the subscripts on the coefficients run counter to the exponents, like this:

$$p(x) = a_0 x^n + a_1 x^{n-1} + \cdots + a_{n-1} x + a_n.$$

The *degree* of a polynomial is denoted $\deg p(x)$; it is the largest exponent on x associated with a nonzero coefficient. If $p(x) = 0$ (the zero polynomial), then $\deg p$ is either undefined or $-\infty$.

The set of all polynomials in the variable x with real coefficients is denoted $\mathbb{R}[x]$. More generally, if R is any ring[2], then $R[x]$ is the set of all polynomials in the variable x with coefficients in R.

Polynomials may have more than one variable, e.g., $x^3 y^2 - 3xy + 2x - 3$. The notation $R[x, y]$ stands for the set of all polynomials with coefficients from R in the variables x and y.

Just as we have the notions of divisibility | and mod for integers, the same notions apply to polynomials:

- $p(x)|q(x)$ means there is a polynomial $a(x)$ so that $a(x)p(x) = q(x)$.
- $p(x) \equiv q(x) \pmod{a(x)}$ means that $a(x)|p(x) - q(x)$.
- $R[x]/(a(x))$ is the set of polynomials (with coefficients from R) taken modulo $a(x)$. For example, $\mathbb{Z}[x]/(x^2 + 1)$ is, effectively, the Gaussian integers.

Certain families of polynomials have their own notation. These include the following:

- **Chebyshev polynomials**. The degree-n Chebyshev polynomial of the first kind is denoted $T_n(x)$ and the degree-n polynomial of the second kind is denoted $U_n(x)$.

---

[2]See footnote on page 19.

- **Hermite polynomials**. The degree-n Hermite polynomial is denoted $H_n(x)$. There are two variations on how these are defined (for the probability community and the physics community).
- **Laguerre Polynomials**. The degree-n Laguerre polynomial is denoted $L_n(x)$.
- **Legendre polynomials**. The degree-n Legendre polynomial is denoted $P_n(x)$.

Power series. A *power series* is a function of the form

$$f(x) = a_0 + a_1 x + a_2 x^2 + \cdots .$$

The set of all power series, with coefficients drawn from a ring R, is denoted $R[\![x]\!]$. These are sometimes called *formal* power series when convergence is not at issue.

Rational functions. A *rational function* is the ratio of two polynomials:

$$f(x) = \frac{a_n x^n + a_{n-1} x^{n-1} + \cdots + a_1 x + a_0}{b_m x^m + b_{m-1} x^{m-1} + \cdots + b_1 x + b_0}.$$

The set of all rational functions with real coefficients is denoted $\mathbb{R}(x)$. In general, if F is any field[3], $F(x)$ is the set of rational functions in the variable x with coefficients drawn from F.

5. Miscellany

Here are a few additional notations concerning functions one might encounter.

- **Arg min and arg max**. Given a function f, we write arg max $f(x)$ to stand for the value of x that makes $f(x)$ as large as possible. Similarly, arg min $f(x)$ is the value of x that minimizes f.
- **Functions as superscripts**. On rare occasions one may see the application of a function to its argument—usually written $f(x)$—expressed by writing the function as an exponent. That is, x^f means $f(x)$.
- **Arrows to denote one-to-one and onto**. Sometimes the arrow in $f : A \to B$ is modified to give special meaning. This can be done by writing words above the arrow, such as

$$f : A \xrightarrow{1:1} B \quad \text{or} \quad f : A \xrightarrow{\text{onto}} B$$

to indicate that f is one-to-one or onto, respectively. However, some people "decorate" the arrow to give the same meanings:

$$f : A \hookrightarrow B \quad \text{or} \quad f : A \twoheadrightarrow B$$

where the hooked arrow means "one-to-one" and the double-headed arrow means "onto".

---

[3]See footnote on page 18.

Incidentally, a one-to-one function is called an *injection* and an onto function is called a *surjection*. While these terms have waning popularity, the term *bijection*—a function that is both an injection and a surjection—is commonly used.

- **Commutative diagrams**. Sometimes one may encounter a figure illustrating multiple functions between various sets with an announcement that "this diagram commutes." For example, the following is such a *commutative diagram*:

$$
\begin{array}{ccccc}
A & \xrightarrow{\alpha} & B & \xrightarrow{\beta} & C \\
\downarrow{f} & & \downarrow{g} & & \downarrow{h} \\
D & \xrightarrow{\gamma} & E & \xrightarrow{\delta} & F
\end{array}
$$

In this diagram, the capital letters stand for sets and the labeled arrows stand for functions. The arrow from A to B labeled α means that α is a function from A to B, i.e., $\alpha : A \to B$. Likewise $h : C \to F$, and so on. The assertion that this diagram *commutes* means that any path from one set to another, formed by composing functions, gives the same result. In this diagram there are several paths from A to F including

$$
A \xrightarrow{\alpha} B \xrightarrow{\beta} C \xrightarrow{h} F \quad \text{and} \quad A \xrightarrow{f} D \xrightarrow{\gamma} E \xrightarrow{\delta} F.
$$

The *commutative* property in this case means that the compositions $h \circ \beta \circ \alpha$ and $\delta \circ \gamma \circ f$ are equal, i.e.,

$$
\forall a \in A, \ (h \circ \beta \circ \alpha)(a) = (\delta \circ \gamma \circ f)(a).
$$

- **Exact sequences**. Another function diagram notation is concept of an *exact sequence* of functions. Suppose $A_0, A_1, \ldots, A_n$ are algebraic structures (e.g., groups, modules, vector spaces) and we have functions $f_1, \ldots, f_n$ arranged like this:

$$
A_0 \xrightarrow{f_1} A_1 \xrightarrow{f_2} A_2 \xrightarrow{f_3} \cdots \xrightarrow{f_n} A_n.
$$

That is $f_i : A_{i-1} \to A_i$. In this algebraic context, it is understood that the functions are homomorphisms.

To say that this sequence is *exact* means $\operatorname{Im} f_i = \ker f_{i+1}$ for all $1 \le i < n$. (The *kernel* of f is the set $\ker f = \{x : f(x) = 0\}$. See page 43.) In other words, for all $i = 1, 2, \ldots, n-1$ and for all $a \in A_{i-1}$, we have $f_{i+1}[f_i(a)] = 0$ (or the identity element of A_{i+1}).

This notion of an exact sequence has two interesting special cases. Suppose we have[4] that

$$
0 \longrightarrow A \xrightarrow{f} B
$$

---

[4]Strictly speaking, the leftmost set in this sequence should be written $\{0\}$.

is exact. The image of the unnamed first function must be the 0-element of A and so the exactness property requires that $\ker f = \{0\}$ which, in turn, implies that f is one-to-one.

On the other hand, suppose

$$A \xrightarrow{\ f\ } B \longrightarrow 0$$

is an exact sequence. This means that the entire space B is mapped to 0 by the unnamed function on the right, and so $\operatorname{Im} f = B$; that is, f must be onto.

CHAPTER 7

Linear Algebra

1. Vectors

A (real) vector is an n-long column[1] of real numbers:

$$\mathbf{x} = \begin{bmatrix} x_1 \\ x_2 \\ \vdots \\ x_n \end{bmatrix}.$$

In textbooks, vectors are often denoted with bold letters $\mathbf{v}$, but some people write vectors as ordinary letters with an arrow on top $\vec{v}$. However, when writing vectors by hand (on paper or on a black board), it's easier to draw a single-tined arrow (called a harpoon): $\vec{v}$. Another convention for handwritten vectors is $\underset{\sim}{v}$ as the wavy underline is a typographer's convention for boldface. Finally, many mathematicians use plain letters v (no boldface or arrows) to stand for vectors.

The set of all vectors of length n is denoted $\mathbb{R}^n$. (Complex vectors are n-long columns of complex numbers, and the set of all such vectors is denoted $\mathbb{C}^n$.)

Vectors corresponding to physical quantities (such as forces or displacements) are typically three dimensional, i.e., are elements of $\mathbb{R}^3$. Such vectors are often expressed using $\mathbf{i}, \mathbf{j}, \mathbf{k}$-coordinates. These basis vectors are defined as

$$\mathbf{i} = \begin{bmatrix} 1 \\ 0 \\ 0 \end{bmatrix}, \quad \mathbf{j} = \begin{bmatrix} 0 \\ 1 \\ 0 \end{bmatrix}, \quad \text{and} \quad \mathbf{k} = \begin{bmatrix} 0 \\ 0 \\ 1 \end{bmatrix}.$$

In $\mathbb{R}^n$ the standard basis vectors are usually denoted $\mathbf{e}_1, \mathbf{e}_2, \ldots, \mathbf{e}_n$ where $\mathbf{e}_j$ is a vector entirely populated by zeros except for a 1 in position j. (The dimension of the vector, n, is inferred from context.)

Whether one writes $\mathbf{i}, \mathbf{j}, \mathbf{k}$ or $\mathbf{e}_1, \mathbf{e}_2, \ldots$, all of these vectors have length one (see entry under *Magnitude* below) and are called *unit vectors*. Some people indicate that a vector has length one with a hat decoration: $\hat{\mathbf{u}}$.

The vector of all zeros is variously denoted $\mathbf{0}$, $\vec{0}$, or sometimes simply as 0 (in which case it is difficult to distinguish from the real number zero). A vector of all ones can be written $\mathbf{1}$, $\vec{1}$, $\mathbf{e}$, or $\vec{e}$. We prefer the notation $\mathbf{1}$.

Standard vector operations:

---

[1]Some people write vectors as row vectors. This is convenient because the notation is more compact. It is also traditional in some application areas.

- **Sum**. The sum of vectors **x** and **y** is simply **x** + **y**.
- **Scalar multiple**. The scalar multiplication is usually written with the scalar s on the left and the vector **x** on the right: s**x**.
- **Magnitude**. The magnitude (length, norm) of a vector is denoted $\|\mathbf{x}\|$ and we have

$$\|\mathbf{x}\| = \sqrt{x_1^2 + x_2^2 + \cdots + x_n^2}.$$

Some people use single bars $|\mathbf{x}|$ to denote the magnitude of a vector.

- **p-norm**. For a positive real number p, the p-norm of **x** is

$$\|\mathbf{x}\|_p = \left[|x_1|^p + |x_2|^p + \cdots + |x_n|^p \right]^{1/p}.$$

Note that $\|\mathbf{x}\|_2 = \|\mathbf{x}\|$.

The infinity norm is $\|\mathbf{x}\|_\infty = \max\{|x_1|, |x_2|, \cdots, |x_n|\}$.

- **Dot (inner) product**. The dot product of two vectors **x** and **y** is

$$\mathbf{x} \cdot \mathbf{y} = x_1 y_1 + x_2 y_2 + \cdots + x_n y_n.$$

The dot product of **x** and **y** is also written $\langle \mathbf{x}, \mathbf{y} \rangle$. These delimiters are called *angle brackets* and are not less-than and greater-than symbols. The dot product may also be written as $\mathbf{x}^t \mathbf{y}$ or $\mathbf{x}^T \mathbf{y}$ (see the discussion of *Transpose* on page 42).

If the angle between vectors **x** and **y** is θ, then we have the formula

$$\mathbf{x} \cdot \mathbf{y} = \|\mathbf{x}\| \times \|\mathbf{y}\| \times \cos \theta.$$

If **x** and **y** are complex n-vectors, then often

$$\langle \mathbf{x}, \mathbf{y} \rangle = \sum_{k=1}^{n} x_k \overline{y_k}$$

but some authors define this as $\sum_{k=1}^{n} \overline{x_k} y_k$. (Here $\overline{x}$ denotes complex conjugate; see page 17.)

Physicists denote inner products like this: $\langle \mathbf{x} \mid \mathbf{y} \rangle$. This is known as *bra-ket* (or *bracket*) notation. They also write $\langle x |$ and $| y \rangle$ for row and column vectors which they call bra-vectors and ket-vectors.

Inner products arise in contexts beyond $\mathbb{R}^n$ or $\mathbb{C}^n$. For example, if $f, g : \mathbb{R} \to \mathbb{R}$, then we may put

$$\langle f, g \rangle = \int_{-\infty}^{\infty} f(t) g(t) \, dt$$

or there may be a weighting in the formula as in this example:

$$\langle f, g \rangle = \int_{-\infty}^{\infty} e^{-t^2/2} f(t) g(t) \, dt.$$

- **Orthogonality**. Two vectors are *orthogonal* provided their dot product is zero, in which case we write $\mathbf{x} \perp \mathbf{y}$.

The $\perp$ symbol is also used to define subspaces of $\mathbb{R}^n$. If $\mathcal{V}$ is a subspace of $\mathbb{R}^n$, then $\mathcal{V}^{\perp}$ is the set of vectors that are orthogonal to all vectors in $\mathcal{V}$:

$$\mathcal{V}^{\perp} = \{\mathbf{x} \in \mathbb{R}^n : \forall \mathbf{v} \in \mathcal{V}, \ \mathbf{x} \perp \mathbf{v}\}.$$

The subspace $\mathcal{V}^{\perp}$ is called the *orthogonal complement* of $\mathcal{V}$.

- **Cross product**. The cross product is defined only for vectors in $\mathbb{R}^3$. If $\mathbf{v} = \begin{bmatrix} v_1 \\ v_2 \\ v_3 \end{bmatrix}$ and $\mathbf{w} = \begin{bmatrix} w_1 \\ w_2 \\ w_3 \end{bmatrix}$ then

$$\mathbf{v} \times \mathbf{w} = \begin{bmatrix} v_2 w_3 - v_3 w_2 \\ v_3 w_1 - v_1 w_3 \\ v_1 w_2 - v_2 w_1 \end{bmatrix}.$$

- **Scalar triple product**. Given three vectors $\mathbf{a}, \mathbf{b}, \mathbf{c} \in \mathbb{R}^3$, their scalar triple product is

$$[\mathbf{a}, \mathbf{b}, \mathbf{c}] = \mathbf{a} \cdot (\mathbf{b} \times \mathbf{c}) = (\mathbf{a} \times \mathbf{b}) \cdot \mathbf{c} = \det \begin{bmatrix} a_1 & a_2 & a_3 \\ b_1 & b_2 & b_3 \\ c_1 & c_2 & c_3 \end{bmatrix}.$$

2. Matrices

A matrix is a rectangular array of numbers. A matrix with m rows and n columns is called an $m \times n$-matrix; the number of rows is always given first and then the number of columns. The set of all real $m \times n$-matrices is denoted $\mathbb{R}^{m \times n}$ and for complex matrices we write $\mathbb{C}^{m \times n}$. An alternative notation is $M_n(\mathbb{R})$ for real $n \times n$-matrices and $M_{a,b}(\mathbb{R})$ for real $a \times b$-matrices. Likewise for complex matrices, we write $M_n(\mathbb{C})$ and $M_{a,b}(\mathbb{C})$.

It is often useful to consider vectors as $n \times 1$-matrices.

If A is a matrix, the i, j-entry of A is the entry in the ith row and j^{th} column and is variously denoted $A_{i,j}$ or $a_{i,j}$ or $[A]_{i,j}$. It is common to omit the comma separating the subscripts.

There are many notations associated with matrices.

- **Addition/subtraction**. Matrices of the same shape may be added. The i, j-entry of $A + B$ is $a_{i,j} + b_{i,j}$. Likewise $A - B$ is the matrix whose i, j-entry is $a_{i,j} - b_{i,j}$.
- **Scalar multiplication**. If A is a matrix and r is a scalar, then rA is a matrix, of the same size as A, whose i, j-entry is $ra_{i,j}$.
- **Matrix multiplication**. If A is an $m \times n$-matrix and B is an $n \times p$-matrix, then they may be multiplied and the i, j-entry of AB is

$$[AB]_{i,j} = \sum_{k=1}^{n} a_{i,k} b_{k,j}.$$

- **Identity matrix and Kronecker's delta**. The *identity matrix* is an $n \times n$-matrix whose off diagonal entries are all 0 and whose main diagonal entries are all 1. An identity matrix is denoted I; the notation I_n is used to denote an $n \times n$-identity matrix.

The notation $\delta(i, j)$, called *Kronecker's delta*, is used to represent the i, j-entry of an identity matrix. It is defined by

$$\delta(i, j) = \begin{cases} 1 & \text{if } i = j \text{ and} \\ 0 & \text{if } i \neq j. \end{cases}$$

Kronecker's delta is sometimes written with its arguments as subscripts: $\delta_{i,j}$ or δ_{ij}.

- **Inverse**. If A is a square matrix and there is another square matrix B such that $AB = I$, then B is called the inverse of A and is denoted A^{-1}.

- **Pseudoinverse**. The Moore-Penrose pseudoinverse of a matrix A is denoted A^+. Some people write $A^\dagger$, but this notation is also used for the conjugate transpose of A (see *Transpose* on the current page).

- **Hadamard product**. Given two matrices A and B of the same shape, their *Hadamard* product is denoted $A \circ B$. It is a new matrix with the same shape as A and B whose i, j-entry is $a_{i,j}b_{i,j}$.

- **Matrix of all 1s**. A matrix of all 1s is often denoted J. We write $J_{m,n}$ to stand for an $m \times n$-matrix of all 1s.

- **Transpose**. If A is an $m \times n$-matrix, its *transpose* is an $n \times m$-matrix denoted A^t (or A^T); it is defined by

$$\left[A^t \right]_{i,j} = a_{j,i}.$$

A matrix A that satisfies $A = A^t$ is called *symmetric*.

For a (complex) $m \times n$-matrix A, its *conjugate transpose* is an $n \times m$-matrix A^*; it is defined by

$$[A^*]_{i,j} = \overline{a_{j,i}}.$$

Some authors prefer an H superscript to denote conjugate transpose: A^H. In addition, some people write $A^\dagger$, but this may be confused with the notation for the Moore-Penrose pseudoinverse (described earlier).

[Note: In the MATLAB programming language, the transpose of a matrix is denoted with a prime symbol: `A'`. Note that if A is a complex matrix, this returns the conjugate transpose of A. If one desires the non-conjugated transpose, then `A.'` should be used.]

The matrix A^* (or A^H) is known as the *adjoint* of A.

A matrix that satisfies $A = A^*$ (equivalently $A = A^H$) is called *Hermitian* or *self-adjoint*.

Note: The adjoint of A should not be confused with the *adjugate* of A, denoted $\text{adj}(A)$, which is the transpose of the matrix of cofactors of A.

- **Trace**. The *trace* of a matrix is the sum of its diagonal elements. If A is an $m \times n$-matrix then

$$\text{tr}\, A = \sum_{i=1}^{\min\{m,n\}} a_{i,i}.$$

- **Rank/nullity**. The *rank* of a matrix A is the dimension of its column (or row) space and is denoted rank(A), though other notations may be used.

  The *null space* of A, the set of vectors $\{\mathbf{x} : A\mathbf{x} = \mathbf{0}\}$, may be denoted null($A$). It is also denoted ker(A) because the null space is also known as the *kernel* of A.

- **Determinant**. The *determinant* of a matrix is denoted $\det A$. Sometimes the determinant of a matrix is indicated by vertical bars in place of square brackets:

$$\begin{vmatrix} 1 & 2 & 3 \\ 4 & 7 & 0 \\ 2 & 1 & 9 \end{vmatrix} = \det \begin{bmatrix} 1 & 2 & 3 \\ 4 & 7 & 0 \\ 2 & 1 & 9 \end{bmatrix}.$$

Some people place the vertical bars around the name of the matrix to denote determinant: $|A|$.

The determinant of an $n \times n$-matrix A can be expressed by the following formula:

$$\det A = \sum_{\pi \in S_n} (\text{sgn}\,\pi) a_{1,\pi(1)} a_{2,\pi(2)} \cdots a_{n,\pi(n)}. \tag{1}$$

This formula affords us the opportunity to discuss additional notation!

- S_n stands for the set of all permutations on the set $\{1, 2, \ldots, n\}$. Thus the sum has $n!$ terms; one for each permutation of $[n]$. It is called the *symmetric group*. Some people use a German S for this set: $\mathfrak{S}_n$.
- π in this formula is not the real number $3.14159\ldots$ but rather is a dummy variable standing for a permutation.
- $\text{sgn}\,\pi$ is the *sign* of the permutation π; it equals 1 if π is an even permutation and -1 if π is an odd permutation.

- **Permanent**. By omitting the $\text{sgn}\,\pi$ factor in the formula in equation (1), we arrive at the *permanent* of the matrix A:

$$\text{perm}\, A = \sum_{\pi \in S_n} a_{1,\pi(1)} a_{2,\pi(2)} \cdots a_{n,\pi(n)}.$$

- **Matrix powers**. Let A be a square matrix. For an integer n,

$$A^n = \begin{cases} \underbrace{A \cdot A \cdots A}_{n \text{ times}} & \text{for } n > 0, \\ I & \text{for } n = 0, \text{ and} \\ \left(A^{-1}\right)^{|n|} & \text{for } n < 0, \text{ provided } A \text{ is invertible.} \end{cases}$$

Given a square matrix A, the *square root* of A should be a matrix B so that $B^2 = A$. In case A is real symmetric or Hermitian, and positive semi-definite, then $\sqrt{A} = A^{1/2}$ has a generally agreed upon interpretation: Diagonalize $A = S^*\Lambda S$ where $S^*S = I$ and put $\sqrt{A} = S^*\sqrt{\Lambda}S$ where $\sqrt{\Lambda}$ is the diagonal matrix of the (nonnegative) square roots of the eigenvalues.

- **Matrix exponential**. Let A be a square matrix. The expression $\exp A$ (or e^A) is the *matrix exponential* function. It means

$$\exp A = \sum_{k=0}^{\infty} \frac{1}{k!}A^k.$$

- **Tensor/Kronecker product/sum**. Let A be an $m \times n$-matrix and B be a $p \times q$-matrix. Their *tensor* (or *Kronecker*) product is an $mp \times nq$-matrix given by the following formula:

$$A \otimes B = \begin{bmatrix} a_{1,1}B & a_{1,2}B & \cdots & a_{1,n}B \\ a_{2,1}B & a_{2,2}B & \cdots & a_{2,n}B \\ \vdots & \vdots & \ddots & \vdots \\ a_{m,1}B & a_{m,2}B & \cdots & a_{m,n}B \end{bmatrix}.$$

For square matrices A and B (where A is $a \times a$ and B is $b \times b$) their *Kronecker sum* is denoted $A \oplus B$ and is defined by

$$A \oplus B = A \otimes I_b + I_a \otimes B$$

where I_a and I_b are identity matrices whose sizes are given by their subscripts. Be warned! The notation $A \oplus B$ has another meaning that we consider next.

- **Direct sum**. For any two matrices A and B, their *direct sum* is denoted $A \oplus B$. This is the block diagonal matrix with the structure

$$\begin{bmatrix} A & 0 \\ 0 & B \end{bmatrix}$$

where the 0's represent blocks of zeros of the appropriate size.

Note that $\oplus$ is also stands for the Kronecker sum (see the previous entry).

- **Inequalities**. If A and B are the same shape, then $A \geq B$ means that each term in A is greater than or equal to the corresponding term in B, i.e., $a_{i,j} \geq b_{i,j}$. Likewise $A > B$, $A < B$, and $A \leq B$ have similar meanings. However, $A \neq B$ simply means that the two matrices are not identical.

  The notation $A \geq 0$ means that every entry in A is nonnegative, and likewise for $A > 0$, $A \leq 0$, and $A < 0$.

  If A and B are square, real symmetric or complex Hermitian matrices, $A \geq B$ means that $A - B$ is positive semidefinite and $A > B$ means that $A - B$ is positive definite. The notations $A \leq B$ and $A < B$ mean $B \geq A$ and $B > A$, respectively. Writing $A > 0$ [resp. $A \geq 0$] means that A is positive definite [resp. positive semidefinite]. Of course, $A < 0$ [$A \leq 0$] means that A is negative definite [negative semidefinite].

- **Eigenvalues**. It is strongly traditional (but not mandatory) to use the letter λ for an eigenvalue of a matrix and subscripted λ's (i.e., $\lambda_1, \ldots, \lambda_n$) to denote the full spectrum.

  If the eigenvalues are all real (as is the case for Hermitian matrices), then one often subscripts the eigenvalues in order, i.e., $\lambda_1 \leq \lambda_2 \leq \cdots \leq \lambda_n$ or the reverse. In either case, we use the notation $\lambda_{\min}$ and $\lambda_{\max}$ for the smallest and largest eigenvalues, respectively.

- **Singular values**. It is traditional to use the letters σ or s to stand for a singular value of a matrix A. The full list of singular values are usually subscripted in numerical order: $\sigma_1 \geq \sigma_2 \geq \cdots \geq \sigma_n$ or the reverse. We write $\sigma_{\max}$ and $\sigma_{\min}$ for the largest and smallest singular values, respectively.

- **Matrix norms**. We use a triple vertical line to denote matrix norms: $\lVert A \rVert$. Note that viewing $\mathbb{C}^{n \times n}$ as a vector space allows us to consider vector norms, but matrix norms require the submultiplicative property: $\lVert AB \rVert \leq \lVert A \rVert \cdot \lVert B \rVert$; the triple delimiter distinguishes true matrix norms from vector norms on $\mathbb{C}^{n \times n}$ viewed merely as an n^2-dimensional subspace. However, some authors use the double bars $\|A\|$ for matrix norms.

  Note: We have not found a satisfactory method for typesetting $\lVert A \rVert$ in LaTeX. The solution we employ is to define commands like this:

```
\newcommand{\ltriple}{\lvert\hskip -1pt\lvert\hskip -1pt\lvert}
\newcommand{\rtriple}{\rvert\hskip -1pt\rvert\hskip -1pt\rvert}
```

  and then type `$\ltriple A \rtriple$` to make $\lVert A \rVert$.

  Here are some specific norms defined on matrices.

  - **Spectral norm**. Denoted $\lVert A \rVert_2$, the *spectral norm* is defined by

$$\lVert A \rVert_2 = \sqrt{\lambda_{\max}(A^*A)} = \sigma_{\max}(A).$$

  Sometimes the subscript 2 is omitted.

- **Maximum absolute column sum norm.** This is denoted $\|A\|_1$ and is defined to be

$$\|A\|_1 = \max_j \sum_i |a_{i,j}|.$$

- **Maximum absolute row sum norm.** This is denoted $\|A\|_\infty$ and is defined to be

$$\|A\|_\infty = \max_i \sum_j |a_{i,j}|.$$

- **Operator norm.** Every vector norm on $\mathbb{C}^n$ induces a matrix norm on $\mathbb{C}^{n \times n}$. In particular, for the p-norms we have this:

$$\|A\|_p = \max_{\mathbf{x}:\|\mathbf{x}\|_p=1} \|A\mathbf{x}\|_p.$$

- **Frobenius norm.** The *Frobenius norm* of A is denoted $\|A\|_F$. It equals

$$\|A\|_F = \left(\sum_{i,j} |a_{i,j}|^2 \right)^{\frac{1}{2}}.$$

It is not a matrix norm.

- **Spectral radius.** The *spectral radius* of a square matrix A is denoted $\rho(A)$. It is defined by

$$\rho(A) = \max\{|\lambda| : \lambda \text{ is an eigenvalue of } A\}.$$

- **Condition number.** The condition number of a matrix A is denoted $\kappa(A)$. It equals $\sigma_{\max}/\sigma_{\min}$.

Certain collections of matrices form a group under matrix multiplication. Here are notations associated with some of these groups.

- **General linear group.** $\mathrm{GL}_n(\mathbb{R})$ is the group of real $n \times n$-matrices that are invertible. The $\mathbb{R}$ can be replaced by $\mathbb{C}$ or other rings/fields.
- **Special linear group.** $\mathrm{SL}_n(\mathbb{R})$ is the group of real $n \times n$-matrices with determinant equal to 1. The $\mathbb{R}$ can be replaced by $\mathbb{C}$ or other rings/fields.
- **Orthogonal group.** $\mathrm{O}(n)$ is the group of real $n \times n$-matrices that are orthogonal, i.e., $U^t U = I$. This may also be written $\mathrm{O}(n, \mathbb{R})$.
  The notation $\mathrm{O}(n, \mathbb{C})$ is the set of all complex $n \times n$-matrices U such that $U^* U = I$.
- **Special orthogonal group.** $\mathrm{SO}(n)$ is the subgroup of $\mathrm{O}(n)$ with the added condition that $\det U = 1$. Likewise for $\mathrm{SO}(n, \mathbb{C})$.

3. Tensors

In its simplest form, a *tensor* is a multidimensional array of numbers. As such, an entry of, say, an order-3 tensor T is $T_{i,j,k}$ or T_{ijk}. (Some authors use the word *rank* instead of *order*.) Thus scalars are order-0 tensors, vectors are order-1, and matrices are order-2. Higher order tensors are typically written with capital letters in either italics T or boldface $\mathbf{T}$.

More abstractly, a tensor is an element of the (repeated) tensor product of a vector space $\mathcal{V}$ with itself and/or its dual space $\mathcal{V}^*$. That is, a type-(n, m) tensor T is an element of

$$\underbrace{\mathcal{V} \otimes \mathcal{V} \otimes \cdots \otimes \mathcal{V}}_{n \text{ times}} \otimes \underbrace{\mathcal{V}^* \otimes \mathcal{V}^* \otimes \cdots \otimes \mathcal{V}^*}_{m \text{ times}}.$$

The order of an (n, m)-tensor is $n + m$.

An entry in an (n, m)-tensor is indicated with n upper indices (written as superscripts) and m lower indices (written as subscripts):

$$T^{i_1, i_2, \dots, i_n}_{j_1, j_2, \dots, j_m}.$$

In particular, Kronecker's delta in tensor notation is often seen with one upper and one lower index:

$$\delta^i_j = \begin{cases} 1 & \text{if } i = j \text{ and} \\ 0 & \text{otherwise.} \end{cases}$$

Einstein notation is often used with tensors. Typically, the repeated index is a lower index in one tensor and an upper index in the other:

$$T^{ij}_k v^k := \sum_k T^{ij}_k v^k.$$

CHAPTER 8

Calculus

1. Limits

The principal notation for limits is

$$\lim_{x \to a} f(x)$$

which stands for the limit of $f(x)$ as x approaches a.

One sided limits have the notation

$$\lim_{x \to a-} f(x) = \lim_{x \uparrow a} f(x)$$

for the limit of $f(x)$ as x approaches a from the left and

$$\lim_{x \to a+} f(x) = \lim_{x \downarrow a} f(x)$$

for the limit of $f(x)$ as x approaches a from the right. We also have limits in which the variable increases or decreases without bound:

$$\lim_{x \to \infty} f(x) \qquad \text{and} \qquad \lim_{x \to -\infty} f(x).$$

For a sequence $a_1, a_2, a_3, \ldots$ we also use the notation

$$\lim_{n \to \infty} a_n$$

to express the limiting value of the sequence. In addition we have

$$\limsup_{n \to \infty} a_n = \lim_{n \to \infty} \left[\sup\{a_k : k \geq n\} \right] \qquad \text{and}$$

$$\liminf_{n \to \infty} a_n = \lim_{n \to \infty} \left[\inf\{a_k : k \geq n\} \right].$$

We may also write $\overline{\lim}$ for lim sup and $\underline{\lim}$ for lim inf.

2. Derivatives (single independent variable, scalar- or vector-valued)

Given a function f, we have two standard notations for the derivative of f: Newton's notation f' and Leibniz's notation $\frac{df}{dx}$. Some authors use a roman (as opposed to italic) d in this notation, like this: dy/dx.

Higher order derivatives in Newton's notation are expressed by using additional prime marks. The second derivative is f'', the third is f''', and so forth. Sometimes lower case roman numerals replace multiple prime marks, e.g. $f^{(iv)}$. For a positive integer n, the nth derivative may be written $f^{(n)}$.

49

Higher order derivatives in Leibeniz's notation are expressed using exponents like this:

$$\frac{df}{dx}, \frac{d^2 f}{dx^2}, \frac{d^3 f}{dx^3}, \dots$$

The nth derivative is written $\frac{d^n f}{dx^n}$.

The value of a derivative at a specific value (say, at $x = a$) is written as $f'(a)$ in Newton's notation. The Leibniz notation is more cumbersome:

$$\left.\frac{df}{dx}\right|_{x=a}$$

The values of the second derivative are written $f''(a)$ and $\left.\frac{d^2 f}{dx^2}\right|_{x=a}$ in the two notation styles, and so forth for higher order derivatives.

Another style of notation for derivative involves placing dots over the function. Suppose y is a function of t (which often represents time). Then $\dot{y}$ denotes the derivative dy/dt. Likewise $\ddot{y}$ is the second derivative $d^2 y/dt^2$.

The notation $\frac{d}{dx}$ means "the derivative of". It is also denoted with a simple capital D:

$$\frac{d}{dx}\left[x^2 - 3x + 2\right] = D\left[x^2 - 3x + 2\right] = 2x - 3.$$

Higher order derivatives are denoted as $\frac{d^n}{dx^n}$ or D^n.

The notations for the derivative of a vector-valued function of a single variable are the same as for that of a single-valued function. Let $f: \mathbb{R} \to \mathbb{R}^n$. Then the derivative of f is denoted either f' or df/dt (or df/dx). Higher order derivatives are, likewise, f'' and $d^2 f/dt^2$, and so forth.

3. Derivatives (multiple independent variables, scalar-valued)

Partial derivatives. When a function depends on two or more variables, the notion of partial derivative comes into play. Suppose $f: \mathbb{R}^n \to \mathbb{R}$. Then the partial derivative of f with respect to its jth argument is

$$\frac{\partial f}{\partial x_j}.$$

When f is a function of (say) just two variables x and y, then the partial derivatives can also be written as $f_x = \partial f/\partial x$ and $f_y = \partial f/\partial y$. Another notation is $\partial_x f$.

The notation D_x means to take the derivative in the x-direction. Thus

$$D_x f = \partial_x f = f_x = \frac{\partial f}{\partial x}.$$

More generally, if $\mathbf{v}$ is a unit vector then $D_{\mathbf{v}}$ is the derivative of f in the $\mathbf{v}$ direction; this is known as a *directional derivative*.

Higher order partial derivatives are denoted like this:

$$\frac{\partial^2 f}{\partial x_i \partial x_j}$$

which means to first take the partial derivative of f with respect to x_j and then the partial derivative of that result with respect to x_i. In symbols:

$$\frac{\partial^2 f}{\partial x_i \partial x_j} = \frac{\partial}{\partial x_i}\left(\frac{\partial f}{\partial x_j}\right).$$

For most functions likely to be encountered in science and engineering the order of differentiation does not matter.

For a function of (say) two variables x and y, the higher order partials can be written like this: $f_{xx}, f_{xy}, f_{yx}, f_{yy}$.

Gradient. The *gradient* of a function $f : \mathbb{R}^n \to \mathbb{R}$ is the vector of partial derivatives:

$$\begin{bmatrix} \frac{\partial f}{\partial x_1} \\ \frac{\partial f}{\partial x_2} \\ \vdots \\ \frac{\partial f}{\partial x_n} \end{bmatrix}.$$

It is denoted either grad f or ∇f.

For a function f of three variables, the gradient can be expressed in $\mathbf{i}, \mathbf{j}, \mathbf{k}$-notation like this:

$$\text{grad } f = \nabla f = f_x \mathbf{i} + f_y \mathbf{j} + f_z \mathbf{k}.$$

Laplacian. The Laplacian operator is denoted with a capital Δ. For $f : \mathbb{R}^n \to \mathbb{R}$ we have

$$\Delta f = \sum_{i=1}^{n} \frac{\partial^2 f}{\partial x_i^2}$$

or more simply for a function of three variables,

$$\Delta f = \frac{\partial^2 f}{\partial x^2} + \frac{\partial^2 f}{\partial y^2} + \frac{\partial^2 f}{\partial z^2}.$$

Hessian. For a function $f : \mathbb{R}^n \to \mathbb{R}$, we can form an $n \times n$-matrix of its partial second derivatives. This matrix is the *Hessian* matrix of f:

$$Hf = \begin{bmatrix} \frac{\partial^2 f}{\partial x_1 \partial x_1} & \frac{\partial^2 f}{\partial x_1 \partial x_2} & \cdots & \frac{\partial^2 f}{\partial x_1 \partial x_n} \\ \frac{\partial^2 f}{\partial x_2 \partial x_1} & \frac{\partial^2 f}{\partial x_2 \partial x_2} & \cdots & \frac{\partial^2 f}{\partial x_2 \partial x_n} \\ \vdots & \vdots & \ddots & \vdots \\ \frac{\partial^2 f}{\partial x_n \partial x_1} & \frac{\partial^2 f}{\partial x_n \partial x_2} & \cdots & \frac{\partial^2 f}{\partial x_n \partial x_n} \end{bmatrix}.$$

4. Derivatives (multiple independent variables, vector-valued)

In this section we consider derivatives of functions $\mathbf{f} : \mathbb{R}^m \to \mathbb{R}^n$. We denote vector-valued functions with bold letters (though this is not mandatory).

Such functions map a vector $\mathbf{x} \in \mathbb{R}^m$ to a vector $\mathbf{y} \in \mathbb{R}^n$. This can be simply written $\mathbf{y} = \mathbf{f}(\mathbf{x})$ or separated into components like this

$$y_1 = f_1(\mathbf{x})$$
$$y_2 = f_2(\mathbf{x})$$
$$\vdots$$
$$y_n = f_n(\mathbf{x})$$

or like this

$$\mathbf{f}(\mathbf{x}) = \begin{bmatrix} f_1(\mathbf{x}) \\ f_2(\mathbf{x}) \\ \vdots \\ f_n(\mathbf{x}) \end{bmatrix}$$

where, of course, $f_j(\mathbf{x})$ denotes the jth component of the vector $\mathbf{f}(\mathbf{x})$.

Partial derivatives. Partial derivatives of $\mathbf{f}$ typically specify both the component function and the variable with respect to which the function is being differentiated:

$$\frac{\partial f_i}{\partial x_j}.$$

Some people express this partial derivative as $f_{i,j}$.

One can take the partial derivative of all components of $\mathbf{f}$ with respect to a single variable:

$$\frac{\partial \mathbf{f}}{\partial x_j} = \begin{bmatrix} \frac{\partial f_1}{\partial x_j} \\ \frac{\partial f_2}{\partial x_j} \\ \vdots \\ \frac{\partial f_n}{\partial x_j} \end{bmatrix}$$

This vector of partial derivatives can be written $f_{,j}$. One needs to be careful with this notation as the comma may be difficult to notice.

Jacobian. The matrix of all partial derivatives is called the *Jacobian* matrix of f and is simply denoted $D\mathbf{f}$:

$$D\mathbf{f} = \begin{bmatrix} \frac{\partial f_1}{\partial x_1} & \frac{\partial f_1}{\partial x_2} & \cdots & \frac{\partial f_1}{\partial x_m} \\ \frac{\partial f_2}{\partial x_1} & \frac{\partial f_2}{\partial x_2} & \cdots & \frac{\partial f_2}{\partial x_m} \\ \vdots & \vdots & \ddots & \vdots \\ \frac{\partial f_n}{\partial x_1} & \frac{\partial f_n}{\partial x_2} & \cdots & \frac{\partial f_n}{\partial x_m} \end{bmatrix} = \begin{bmatrix} f_{1,1} & f_{1,2} & \cdots & f_{1,m} \\ f_{2,1} & f_{2,2} & \cdots & f_{2,m} \\ \vdots & \vdots & \ddots & \vdots \\ f_{n,1} & f_{n,2} & \cdots & f_{n,m} \end{bmatrix}.$$

If $\mathbf{f} : \mathbb{R}^n \to \mathbb{R}^n$, then $D\mathbf{f}$ is a square matrix; its determinant is, unfortunately, also called the *Jacobian* of $\mathbf{f}$.

One can think of $\mathbf{f} : \mathbb{R}^n \to \mathbb{R}^n$ as a change of coordinate systems. If the original coordinates are, say, (x, y, z), then $\mathbf{f}(x, y, z) = (u, v, w)$ are the co-ordinates of the point in the alternative system. In this way, u, v, and w are themselves functions of x, y, and z. Then the (determinant of the) Jacobian of $\mathbf{f}$ is also denoted:

$$\left| \frac{\partial(u, v, w)}{\partial(x, y, z)} \right| = \det \begin{bmatrix} \frac{\partial u}{\partial x} & \frac{\partial u}{\partial y} & \frac{\partial u}{\partial z} \\ \frac{\partial v}{\partial x} & \frac{\partial v}{\partial y} & \frac{\partial v}{\partial z} \\ \frac{\partial w}{\partial x} & \frac{\partial w}{\partial y} & \frac{\partial w}{\partial z} \end{bmatrix}.$$

Curl. For a function $\mathbf{f} : \mathbb{R}^3 \to \mathbb{R}^3$ the *curl* of $\mathbf{f}$ is

$$\operatorname{curl} \mathbf{f} = \nabla \times \mathbf{f} = \begin{bmatrix} f_{3,2} - f_{2,3} \\ f_{1,3} - f_{3,1} \\ f_{2,1} - f_{1,2} \end{bmatrix}$$

where $f_{1,2}$ means the partial derivative of f_1 in its second argument, i.e., $\partial f_1 / \partial x_2$.

The notation $\nabla \times \mathbf{f}$ is inspired by the cross product in which the first "vector" is a list of operators:

$$\nabla \times \mathbf{f} = \begin{bmatrix} \frac{\partial}{\partial x_1} \\ \frac{\partial}{\partial x_2} \\ \frac{\partial}{\partial x_3} \end{bmatrix} \times \begin{bmatrix} f_1 \\ f_2 \\ f_3 \end{bmatrix}.$$

Divergence. For a function $\mathbf{f} : \mathbb{R}^3 \to \mathbb{R}^3$, the *divergence* of $\mathbf{f}$ is

$$\operatorname{div} \mathbf{f} = \nabla \cdot \mathbf{f} = \frac{\partial f_1}{\partial x_1} + \frac{\partial f_2}{\partial x_2} + \frac{\partial f_3}{\partial x_3}.$$

Like the notation for curl, this notation is inspired by the dot product in which the first "vector" is a list of operators:

$$\nabla \cdot \mathbf{f} = \begin{bmatrix} \frac{\partial}{\partial x_1} \\ \frac{\partial}{\partial x_2} \\ \frac{\partial}{\partial x_3} \end{bmatrix} \cdot \begin{bmatrix} f_1 \\ f_2 \\ f_3 \end{bmatrix}.$$

More generally, for $\mathbf{f} : \mathbb{R}^n \to \mathbb{R}^n$,

$$\operatorname{div} \mathbf{f} = \nabla \cdot \mathbf{f} = \sum_{j=1}^{n} \frac{\partial f_j}{\partial x_j}.$$

Note that for the scalar valued function $f : \mathbb{R}^n \to \mathbb{R}$, the gradient of f, denoted ∇f, is the vector of f's partial derivatives, and so $\nabla f : \mathbb{R}^n \to \mathbb{R}^n$. If we take the divergence of ∇f we have

$$\nabla \cdot (\nabla f) = \sum_{j=0}^{n} \frac{\partial^2 f_j}{\partial x_j^2}$$

which is the Laplacian of f. Thus $\nabla \cdot (\nabla f) = \nabla^2 f$.

5. Integration

Let $f : \mathbb{R} \to \mathbb{R}$. The notation $\int f(x)\,dx$ is the *indefinite integral* of f; it is a function F whose derivative is f. The function F is called an *antiderivative* or a *primitive* of f. The notation

$$\int_a^b f(x)\,dx$$

is the *definite integral* of f and its value can be expressed

$$F(x)\Big|_a^b = F(b) - F(a)$$

where F is an antiderivative of f. Some authors use a roman (as opposed to italic) d in the integral, like this: $\int f(x)\,\mathrm{d}x$.

The left and right end points of the interval over which a function is integrated may be $-\infty$ and ∞, respectively.

Multiple integrals take the following form:

$$\iint f(x,y)\,dxdy.$$

This means we integrate $f(x,y)$ first with respect to x (holding y constant), and then integrate the resulting expression with respect to y. It is equivalent to

$$\int \left[\int f(x,y)\,dx \right] dy.$$

This is not equivalent to $\iint f(x,y)\,dydx$, although the end result is often the same.

Subscripts on integrals indicate the domain of integration. The typical notation is

$$\int_A f\,ds$$

where A is a set. Some examples:

$$\int_{[0,1]} x^2\,dx \text{ means } \int_0^1 x^2\,dx$$

$$\int_{\mathbb{R}} \exp\{-x^2\}\,dx \text{ means } \int_{-\infty}^{\infty} e^{-x^2}\,dx$$

$$\int_{[0,1]^2} (x-y)^2\,dx\,dy \text{ means } \int_0^1 \int_0^1 (x-y)^2\,dx\,dy$$

When the subscript on the integral is a curve

$$\int_\gamma f\,ds$$

denotes the *line integral* of f along the curve γ.

If the curve is closed (i.e., it begins and ends at the same point) then we write

$$\oint_\gamma f \, ds$$

for the line integral. (The subscript indicating the name of the curve may be omitted if it is clear from context.) In this case, the line integral may also be called a *contour integral*. When the contour is the boundary of a domain D, the following notation may be used:

$$\int_{\partial D} f \, ds \quad \text{or} \quad \oint_{\partial D} f \, ds$$

where ∂D indicates the boundary of D.

6. Convolution and transforms

If f and g are functions from $\mathbb{R}$ to $\mathbb{R}$, we define their *convolution* $f * g$ as a new function with

$$(f * g)(x) = \int_{-\infty}^{\infty} f(t)g(x - t) \, dt.$$

The convolution integral need not be computed over the entire real line. Sometimes, it is computed just on an interval especially if the functions f and g are periodic[1]. The interval over which we integrate can be arbitrary:

$$(f * g)(x) = \int_{a}^{b} f(t)g(x - t) \, dt.$$

Common intervals include $[0, 1]$, $[-\pi, \pi]$, and $[0, 2\pi]$.

In some cases, we normalize the integral by dividing by the length of the interval:

$$(f * g)(x) = \frac{1}{b - a} \int_{a}^{b} f(t)g(x - t) \, dt.$$

The notion of convolution (and the use of the symbol $*$) extends to other realms. For example, given sequences $a = (a_0, a_1, a_2, \ldots)$ and $b = (b_0, b_1, b_2, \ldots)$, their convolution is a new sequence $c = a * b$ with

$$c_n = \sum_{k=0}^{n} a_k b_{n-k}.$$

Similarly, in number theory, the convolution of two functions $f, g : \mathbb{Z}^+ \to \mathbb{R}$ is $f * g$ with

$$(f * g)(n) = \sum_{d \mid n} f(d)g(n/d).$$

---

[1]This means there is a positive number p such that $\forall x \in \mathbb{R}, \ f(x) = f(x + p)$.

Let $f : \mathbb{R}_+ \to \mathbb{R}$. The *Laplace transform* of f is denoted $\mathcal{L}f$. It is a new function defined on the nonnegative real numbers by

$$(\mathcal{L}f)(s) = \int_0^\infty f(t)e^{-st}\,dt.$$

The inverse Laplace transform is denoted $\mathcal{L}^{-1}$.

Let $f : \mathbb{R} \to \mathbb{R}$ (or, more generally, $f : \mathbb{R} \to \mathbb{C}$). The *Fourier transform* of f is denoted $\mathcal{F}f$. This is a new function $F : \mathbb{R} \to \mathbb{C}$. Definitions for this vary depending on the author, but the one we prefer is this:

$$(\mathcal{F}f)(\omega) = F(\omega) = \frac{1}{\sqrt{2\pi}} \int_{-\infty}^\infty f(x)\exp\{-i\omega x\}\,dx.$$

Variations on this definition have different coefficients outside the integral and/or omit the minus sign in the exponential term.

The inverse Fourier transform is denoted $\mathcal{F}^{-1}$.

The Fourier transform applies to a function defined on $\mathbb{R}$ while the *discrete Fourier transform* applies to a finite sequence of numbers $a = (a_0, a_1, \ldots, a_{n-1})$. It is also denoted $\mathcal{F}(a)$ and is a new sequence $A = (A_0, A_1, \ldots, A_{n-1})$ with

$$A_k = \sum_{j=0}^{n-1} a_j e^{-jki\pi/n}.$$

Some authors scale this by a factor of $1/\sqrt{n}$ (which we think is preferable). Discrete Fourier transforms can be efficiently computed using the *fast Fourier transform* algorithm, commonly abbreviated FFT.

The discrete Fourier transform is closely related to the *discrete cosine transform* which is commonly abbreviated DCT.

Closely related to the Fourier transform is the *Fourier series* of a (periodic) function. Let $f : [-\pi, \pi] \to \mathbb{R}$. We seek to represent f as a sum of sine and cosine terms like this:

$$f(t) = a_0 + \sum_{k=1}^\infty (a_k \cos kt + b_k \sin kt).$$

This can be written in exponential form:

$$f(t) = \sum_{k=-\infty}^\infty \hat{f}(k)e^{ikt}$$

and the numbers $\hat{f}(k)$ are the *Fourier coefficients*. As with the Fourier transform, variations in the definition exist to account for intervals other than $[-\pi, \pi]$ and with different scaling factors.

CHAPTER 9

Probability and Statistics

1. Probability

Events. In probability theory, one begins with a set of fundamental outcomes, often denoted Ω or S. An *event* is a subset of Ω.

The probability of an event[1] A is typically denoted $P(A)$ but the following notations are also used: $\Pr(A)$ and $\mathbb{P}(A)$.

Given two events, A and B, we have the following notations:

- $P(A \cap B)$ is the probability of A and B. It is also sometimes denoted $P(AB)$ and $P(A \wedge B)$.
- $P(A \cup B)$ is the probability of A or B. It is also sometimes denoted $P(A \vee B)$.
- $P(A \mid B)$ is the *conditional* probability of A given B.

Random variables. Random variables (often abbreviated rv) are generally denoted with capital letters, X. The notation $X(\omega)$ is the value of X at the point ω of the underlying probability space on which X is defined. Notation such as $P(X \geq 0)$ is a shorthand for $P\{\omega : X(\omega) \geq 0\}$.

Given an event A, the notation 1_A is an *indicator* random variable defined by

$$1_A(\omega) = \begin{cases} 1 & \text{if } \omega \in A \text{ and} \\ 0 & \text{otherwise.} \end{cases}$$

Some authors write I_A for indicator random variables.

The following are fundamental notations associated with random variables.

- **Expected value**. For a random variable X, its *expected value* is $E(X)$. Sometimes a different style E is used, e.g., $\mathrm{E}(X)$ or $\mathbb{E}(X)$.

  The letter μ (standing for *mean*) is often used for the expected value of a random variable. If more than one random variable is under consideration, one may subscript μ with the name of that random variable; that is, μ_X is the expected value of X.

  In some disciplines, the notation $\langle X \rangle$ is used to denote the expected value of X.

---

[1]More formally and abstractly, a *probability space* is a triple $(\Omega, \mathcal{F}, P)$ where Ω is a set, $\mathcal{F}$ is a σ-algebra of events, and P is the probability function. In case Ω is finite or countably infinite, the probability space is called *discrete* and $\mathcal{F}$ can be ignored (as it can be taken to be 2^Ω).

- **Conditional expectation.** $E(X \mid Y)$ denotes the conditional expected value of X given Y. This is also written $\mu_{X|Y}$.
- **Variance.** For a random variable X, $\mathrm{Var}(X)$ denotes the *variance* of X defined by by $\mathrm{Var}(X) = E\left[(X - \mu)^2\right] = E(X^2) - E(X)^2$. Often σ^2 denotes the variance. This may be subscripted with the name of the random variable: σ_X^2.

  The square root of the variance is known as the *standard deviation* and is denoted σ.
- **Covariance.** Given two random variables X and Y, their *covariance* is $\mathrm{Cov}(X, Y) = E(XY) - E(X)E(Y)$.

  If **X** is a random vector, i.e.,

$$\mathbf{X} = \begin{bmatrix} X_1 \\ X_2 \\ \vdots \\ X_n \end{bmatrix}$$

  then the *covariance matrix* of **X** is often denoted Σ. This is an $n \times n$-matrix whose i, j-entry is $\mathrm{Cov}(X_i, X_j)$.
- **Correlation coefficient.** Given two random variables X and Y, their *correlation coefficient* is

$$\mathrm{Corr}(X, Y) = \frac{\mathrm{Cov}(X, Y)}{\sigma_X \sigma_Y}.$$

- **Entropy.** The *entropy* of a random variable X is denoted $H(X)$. For discrete random variables,

$$H(X) = -\sum_x P(X = x) \log P(X = x)$$

  where the sum is over all values x that X may take (and with the convention $0 \log 0 = 0$). For a continuous random variable,

$$H(X) = -\int_{-\infty}^{\infty} f(x) \log f(x) \, dx$$

  where $f(x)$ is the PDF (see the entry on the facing page) of X. In both cases, the base of the logarithm depends on the application (with the most popular choices being 2, e, and 10).

  Extensions of this notation include the *joint* entropy $H(X, Y)$ and the *conditional* entropy $H(X \mid Y)$ of the two random variables X and Y.

Distributions. Random variables are often best described by their *distribution*. Distributions, in turn, can be described in one of three principal ways, each with its own acronym.

- **PMF**: Discrete[2] random variables can be specified by their *probability mass function*, or PMF, which specifies the probability associated with each possible value of the random variable: $F(x) = P(X = x)$.
- **CDF**: Real-valued, continuous random variables can be specified by their *cumulative distribution function*, or CDF: $F(x) = P(X \leq x)$. Most authors denoted cumulative distribution functions with upper-case letters.

  The inverse function of the CDF is the *quantile function* which is often denoted Q.
- **PDF**: Real-valued, continuous random variables can also be specified by their *probability density function* $f(x)$. The relation to the cumulative distribution function $F(x)$ is $f(x) = F'(x)$ which gives

$$P(X \in [a, b]) = \int_a^b f(x)\, dx.$$

Most authors use lower case letters that correspond to the upper case letter for the random variable's CDF.

We write $X \sim \mathcal{D}$ if the random variable X has distribution $\mathcal{D}$. For example, if X is a (standard) normal random variable, we write $X \sim N(0, 1)$. (See the discussion of the normal distribution on the next page.)

A collection of random variables with the same distribution are called *identically distributed*. In particular, if a collection of random variables are (mutually) independent and identically distributed, we write that they are iid (or IID).

There are many standard distributions of random variables that have been studied by mathematicians and arise in a host of applications. The following are among the best-known and each have their own notation.

- **Binomial**. Let n be a nonnegative integer and $0 \leq p \leq 1$. A discrete random variable X has the *binomial distribution* $\text{Bin}(n, p)$ or $B(n, p)$ if its probability mass function is

$$P(X = k) = \binom{n}{k} p^k (1 - p)^{n-k}$$

where $k \in \{0, 1, \ldots, n\}$.
- **Poisson**. Let λ be a positive real number. A discrete random variable X has *Poisson distribution* $\text{Pois}(\lambda)$ provided

$$P(X = k) = \frac{\lambda^k}{k!} e^{-\lambda}.$$

where k is a nonnegative integer.

  The use of λ for the parameter (and hence the mean) of the Poisson distribution is quite common.

---

[2]*Discrete* random variables take values in a finite or countably infinite set. *Continuous* random variables generally yield values in all of $\mathbb{R}$ or an interval in $\mathbb{R}$.

- **Exponential distribution**. For a positive real number α, the exponential distribution is denoted $\text{Exp}(\alpha)$. The CDF for a random variable with this distribution is $F(t) = 1 - \exp\{-\alpha t\}$ (where $t \geq 0$).
- **Normal/Gaussian**. A *standard normal* or *Gaussian* random variable X has a probability density function typically denoted $\phi(x)$ where

$$\phi(x) = \frac{1}{\sqrt{2\pi}} \exp\{-x^2\}.$$

The corresponding CDF is often denoted $\Phi(x)$ which is closely related to the error function $\text{erf}(x)$ (see page 28).

More generally, if μ is a real number and σ^2 is a positive real number, then $N(\mu, \sigma^2)$, or sometimes $\mathcal{N}(\mu, \sigma^2)$, is the distribution of a normal random variable X with mean μ and variance σ^2. Its probability density function is

$$\frac{1}{\sqrt{2\pi\sigma^2}} \exp\left\{-\frac{(x-\mu)^2}{2\sigma^2}\right\}.$$

Thus a standard normal random variable has $X \sim N(0, 1)$.

The multivariable normal random variable $N_k(\vec{\mu}, \Sigma)$ is a random k-vector with mean $\vec{\mu}$ and covariance matrix Σ.

- **Chi-squared**. The *chi-squared* distribution is denoted χ^2; more specifically, the chi-square distribution with k-degrees of freedom is denoted χ_k^2. If $X_1, X_2, \ldots, X_k$ are IID standard normal random variables, then the sum of their squares $X = X_1^2 + \cdots + X_k^2$ is chi-squared distributed χ_k^2. Its probability density function is given by

$$\frac{1}{2^{k/2}\Gamma(k/2)} x^{\frac{k}{2}-1} e^{-x/2}$$

where Γ is the gamma function described on page 28.

- **Student's t-distribution**. This distribution is denoted t_ν where ν is a positive integer. It is given by $Z/\sqrt{V/\nu}$ where Z and V are independent random variables, Z is a standard normal, and V is χ_ν^2.
- **F-distribution**. Also known as the Fisher-Snedecor distribution, this is written $F(\nu_1, \nu_2)$ where the ν_i are positive integers. It is given by

$$\frac{X_1/\nu_1}{X_2/\nu_2}$$

where X_1, X_2 are independent χ^2-random variables with ν_1, ν_2 degrees of freedom, respectively.

Convergence of random variables. Let X and $X_1, X_2, \ldots$ be random variables. There are various notions of the sequence X_i converging to X, each with its own notation.

- **Convergence in distribution (in law)**. This is denoted in various ways including these:

$$X_n \xrightarrow{d} X \qquad\qquad X_n \xrightarrow{\mathcal{D}} X$$

$$X_n \xrightarrow{\mathcal{L}} X \qquad\qquad \mathcal{L}(X_n) \longrightarrow \mathcal{L}(X)$$

- **Convergence in probability**. This is denoted

$$X_n \xrightarrow{p} X \qquad \text{or} \qquad X_n \xrightarrow{P} X.$$

- **Almost sure convergence**. This is denoted

$$X_n \xrightarrow{\text{a.s.}} X.$$

- L^2 **convergence**. This is denoted

$$X_n \xrightarrow{L^2} X.$$

Some authors place the notation as a subscript to the arrow rather than above, like this: $X_n \to_d X$.

2. Statistics

The following is a potpourri of notation one encounters in statistics. Given numerical data $X_1, X_2, \ldots, X_n$ we have the following notations:

- **Average**. The (sample) *average* is denoted $\bar{X}$:

$$\bar{X} = \frac{1}{n} \sum_{k=1}^{n} X_k.$$

- **Median**. The (sample) *median* is denoted $\tilde{X}$ or $\check{X}$.
- **Standard Deviation**. The (sample) *standard deviation* is denoted s:

$$s = \sqrt{\frac{\sum (X_i - \bar{X})^2}{n-1}}.$$

- z-**score**. A common way to standardize data is express it as the number of standard deviations above (or below) the mean. For this one uses the z-score:

$$Z_i = \frac{X_i - \mu}{\sigma}$$

where μ is the population mean and σ is the population standard deviation.
- **Order statistics**. The notation $X_{(1)}, X_{(2)}, \ldots, X_{(n)}$ represents the same values but in nondecreasing order. That is $X_{(1)}$ is the smallest X_i, $X_{(n)}$ is the largest, and in general

$$X_{(1)} \le X_{(2)} \le \cdots \le X_{(n)}.$$

In hypothesis testing, the *null hypothesis* is typically denoted H_0 and the *alternative hypothesis* by H_A or H_1. Incorrectly rejecting H_0 when, in fact, it is true is known as a *type I error* and the probability of committing such a mistake is typically denoted by α. Accepting the null hypothesis when, in fact, it is false is known as a *type II error* and the probability of making this mistake is typically denoted β.

In statistical models, error (or noise) terms are often denoted with the letter e or ϵ; these can be subscripted if there are multiple error terms.

Often one is trying to determine some numerical quantity for a population (e.g., the proportion of a population that watch a certain television show) and this is done by computing an estimate from a random sample. It is conventional to place a hat on the estimator of the true parameter; that is, if θ is the parameter whose value we seek, then $\hat{\theta}$ is the estimator of θ.

CHAPTER 10

Approximation

1. Approximate equality of numbers

There are many notations used to capture the concept of being approximately equal. The simplest, and in our opinion the best, is $\approx$. For example, $\pi \approx 3.14$. People use a variety of other symbols for approximate equality including $\doteq$, $\cong$, $\sim$, and $\simeq$.

A more precise way to express approximate equality is to write $\pi = 3.14 \pm 0.002$ which is shorthand for

$$3.14 - 0.002 \leq \pi \leq 3.14 + 0.002.$$

2. Asymptotic relations

A variety of notations exist to show the approximate equality of algebraic expressions (functions), to abbreviate "unimportant" terms, and to measure the quality of the approximation.

For example, one may write: when x is small, $e^x \sim 1 + x + \frac{1}{2}x^2$ or when n is large, $\binom{n}{3} \sim n^3/6$. The symbol $\sim$ means *is asymptotic to* and asserts that the limit of the ratio of the two expressions approaches 1.

To be precise, suppose f, g and functions of x.

- When x is small, $f(x) \sim g(x)$ means

$$\lim_{x \to 0} \frac{f(x)}{g(x)} = 1.$$

- When x is large, $f(x) \sim g(x)$ means

$$\lim_{x \to \infty} \frac{f(x)}{g(x)} = 1.$$

The expression $f(x) \propto g(x)$ indicates that the two functions are *proportional* to each other. Strictly speaking, this means there is a positive number k such that $f(x) = kg(x)$. However, $\propto$ can also be used in the approximate sense in which lower order terms are neglected, e.g., $\binom{n}{3} \propto n^3$. In this case $f(x) \propto g(x)$ means that $f(x)/g(x)$ tends to a finite, nonzero limit as $x \to \infty$.

The symbol $\asymp$ expresses a similar, but even less restrictive, notation of approximate equality. The ratio between the two expressions need not tend to a limit; all that is required is that the ratio be bounded away from 0 and ∞.

More precisely, let $f, g : \mathbb{R} \to \mathbb{R}_+$. The expression $f \asymp g$ means that there are positive numbers t, a, and b such that for all $x \geq t$, $a \leq f(x)/g(x) \leq b$. Using the notation we have discussed earlier:

$$\exists t, a, b > 0, \ \forall x \geq t, \ a \leq \frac{f(x)}{g(x)} \leq b.$$

This can also be expressed: when x is sufficiently large, $a \leq \frac{f(x)}{g(x)} \leq b$.

In some cases, the expression $f(x) \asymp g(x)$ may be applied to functions that return negative values. In that case, we mean that $a \leq |f(x)/g(x)| \leq b$ when x is sufficiently large.

3. Big-oh notation and its relatives

Big oh. The big-oh notation of Landau is a useful shorthand for giving an approximate bound for a function or for indicating (relatively) unimportant terms in an expression.

For example, $\binom{n}{3} = \frac{1}{6}n^3 - \frac{1}{2}n^2 + \frac{1}{3}n$. However, the behavior of $\binom{n}{3}$ for n large is dominated by the first term, $\frac{1}{6}n^3$. The remaining terms are "on the order of" n^2 (or smaller). Thus we may write $\binom{n}{3} = \frac{1}{6}n^3 + O(n^2)$. We may also write $\binom{n}{3} = O(n^3)$ to indicate that $\binom{n}{3}$ is "of the order" n^3 (or smaller).

Another example: For x near 0, e^x is approximately 1. More precisely, $e^x = 1 + x + O(x^2)$. The remaining terms are "of the order" x^2 (or smaller).

The meaning of the notation $f(x) = O(g(x))$ depends on the context: is x near zero or is x large?

- In the case of x near 0, the notation $f(x) = O(g(x))$ means there are positive numbers ϵ and b such that for all x between $-\epsilon$ and ϵ we have

$$\left| \frac{f(x)}{g(x)} \right| \leq b.$$

  More informally, we say: when x is sufficiently small, $|f(x)/g(x)|$ is bounded by b.

- In the case of large x, the notation $f(x) = O(g(x))$ means something similar: there are positive numbers t and b so that for all $x \geq t$ we have

$$\left| \frac{f(x)}{g(x)} \right| \leq b.$$

An expression such as $e^x = 1 + x + O(x^2)$ means $e^x - 1 - x = O(x^2)$.

One rarely (if ever) sees the big oh on the left side of an equation.

An equation of the form $\binom{n}{3} = O(n^3)$ is mildly illogical. The equal sign ought to mean that the two surrounding expressions are the same, and in this case they are not really the same thing. It is more proper to write $\binom{n}{3} \in O(n^3)$

and understand $O(n^3)$ as the set of all functions that are asymptotically bounded by n^3. In practice, most people use the equal sign and tacitly understand not to interpret the equal sign in a strict sense.

Some people use a fancy, calligraphic $\mathcal{O}$ for the big-oh notation.

Big omega and theta. The notation $f(x) = O(g(x))$ is an implicit *upper* bound on $|f(x)|$. It asserts[1] that $|f(x)| \leq b|g(x)|$ where b is some positive number. Thus it is correct to write $\binom{n}{3} = O(n^4)$.

The notation $f(x) = \Omega(g(x))$ is a *lower* bound on $f(x)$. The precise meaning depends on whether we are considering small or large values of x:

- In the case of x near 0, the notation $f(x) = \Omega(g(x))$ means there are positive numbers ϵ and b such that for all x between $-\epsilon$ and ϵ we have
$$\left|\frac{f(x)}{g(x)}\right| \geq b.$$

- In the case of large x, the notation $f(x) = \Omega(g(x))$ means something similar: there are positive numbers t and b so that for all $x \geq t$ we have
$$\left|\frac{f(x)}{g(x)}\right| \geq b.$$

Because Ω represents a lower bound, it is correct to write $\binom{n}{3} = \Omega(n^2)$.

Finally, the notation $f(x) = \Theta(g(x))$ means that both $f(x) = O(g(x))$ and $f(x) = \Omega(g(x))$ hold. That is, there are positive constants a and b so that for all sufficiently large (or small) x we have
$$a \leq \left|\frac{f(x)}{g(x)}\right| \leq b.$$

Note that
$$f(x) = \Theta(g(x)) \iff f(x) \asymp g(x) \iff g(x) = \Theta(f(x)).$$

Thus it is correct to write $\binom{n}{3} = \Theta(n^3)$ but it is incorrect to write $\binom{n}{3} = \Theta(n^k)$ for any $k \neq 3$.

Little oh and omega. The notation $f(x) = o(g(x))$ indicates that $f(x)$ is very much smaller than $g(x)$ in the sense that their ratio tends to zero. As with other notation we've seen, the precise meaning depends on context: small x or large x. In the appropriate case, $f(x) = o(g(x))$ means one of the following:
$$\lim_{x \to 0} \frac{f(x)}{g(x)} = 0 \qquad \text{or} \qquad \lim_{x \to \infty} \frac{f(x)}{g(x)} = 0.$$

For example, when x is small, $\cos x = 1 - \frac{1}{6}x^3 + o(x^4)$, and when n is large, $n! = o(n^n)$.

A $o(1)$ term in an expression stands for a quantity that tends to 0.

---

[1] When x is sufficiently large or small, depending on context.

The notation $f(x) = \omega(g(x))$ has the opposite meaning; that is, $f(x)$ is much larger than $g(x)$. Concisely,

$$f(x) = \omega(g(x)) \iff g(x) = o(f(x)).$$

In other words, depending on context, $f(x) = \omega(g(x))$ means one of the following:

$$\lim_{x \to 0} \left| \frac{f(x)}{g(x)} \right| = \infty \qquad \text{or} \qquad \lim_{x \to \infty} \left| \frac{f(x)}{g(x)} \right| = \infty.$$

In an expression, $\omega(1)$ represents a term that is tending to infinity.

Alternative ways to express the relations $f(x) = o(g(x))$ and $f(x) = \omega(g(x))$ are, respectively, $f(x) \ll g(x)$ and $f(x) \gg g(x)$. The relation $\ll$ expresses the notion of *much less than* and $\gg$ indicates *much greater than*.

For example, for n large, we have $n^5 \ll 2^n$ and $n^n \gg n!$ because

$$\lim_{n \to \infty} \frac{n^5}{2^n} = 0 \qquad \text{and} \qquad \lim_{n \to \infty} \frac{n^n}{n!} = \infty.$$

Bibliography

[1] Florian Cajori. *A History of Mathematical Notations: Two Volumes Bound As One*. Dover, 1993. Originally published in separate volumes by The Open Court Publishing Company in 1928 and 1929.

[2] Michael Downes. Short math guide to LaTeX. `ftp://ftp.ams.org/pub/tex/doc/amsmath/short-math-guide.pdf`.

[3] Michel Goossens, Johannes Braams, and David Carlisle. *The LaTeX Companion: Tools and Techniques for Computer Typesetting*. Addison-Wesley Professional, second edition, 2004.

[4] William Anthony Granville. *Elements of the Differential and Integral Calculus*. Ginn and Company, revised edition, 1911.

[5] George Grätzer. *Math into LaTeX*. Birkhäuser, 2000.

[6] George Grätzer. *More Math into LaTeX*. Birkhäuser, fourth edition, 2007.

[7] Roger A. Horn and Charles R. Johnson. *Matrix Analysis*. Cambridge University Press, 1990.

[8] The Institute of Electrical and Electronics Engineers. *American National Standard Mathematical Signs and Symbols for Use in Physical Sciences and Technology*, 1993.

[9] International Organization for Standardization. *Quantities and Units Part 2: Mathematical Signs and Symbols to be Used in the Natural Sciences and Technology*, 2009. ISO 80000–2:2009, which supercedes ISO 31–11:1992.

[10] Jeff Miller. Earliest uses of various mathematical symbols. `http://jeff560.tripod.com/mathsym.html`.

[11] Scott Pakin. The comprehensive LaTeX symbol list. `http://www.ctan.org/tex-archive/info/symbols/comprehensive/`.

[12] Ambler Thompson and Barry N. Taylor. *Guide for the Use of the International System of Units (SI)*. National Institute of Standards and Technology, 2008. Available online at `http://physics.nist.gov/cuu/pdf/sp811.pdf`.

[13] Eric Weisstein. *Wolfram Mathworld*. `http://mathworld.wolfram.com/`.

[14] Wikipedia. Table of mathematical symbols. `http://en.wikipedia.org/wiki/Table_of_mathematical_symbols`.

Chart

The following table is an index of all the notation we have considered. We give both the page numbers on which the symbols are discussed and the LaTeX commands[2] for producing those symbols.

Symbol	Page(s)	LaTeX
Decorated letters		
a', a''	3, 49	`a',a''` (do not use a\prime)
$a^{(\text{iv})}$	3, 49	`a^{\text{(iv)}}`
$\bar{a}$	3	`\bar{a}`
$\overline{A}$	7, 13, 17	`\overline{A}`
$\hat{a}$	3, 56, 62	`\hat{a}`
$\underset{\sim}{a}$	3, 39	`\utilde{a}`[3]
$\vec{a}, \overrightharp{a}$	39	`\vec{a}`, `\overrightharp{a}`[4]
$\overleftrightarrow{AB}, \overrightarrow{AB}, \overline{AB}$	21	`\overleftrightarrow{AB}`, `\overrightarrow{AB}`, `\overline{AB}`
A^*, A^H	42	`A^*`, `A^H`
A^+	42	`A^+`
$A^\dagger$ (as A^* or A^+)	42	`A^\dagger`
$\dot{y}, \ddot{y}$	50	`\dot y`, `\ddot y`
$\mathbf{0}, \vec{0}$	39	`\mathbf{0}`, `\vec 0`
$\mathbf{1}, \vec{1}$	39	`\mathbf{1}`, `\vec 1`
$\mathbf{1}_A$	33	`\mathbf{1}_A`
Delimiters and punctuation		
$\{\,\}$	5, 28	`\{ \}`
$\{x \in A : \text{condition}\}$	5	`\{x \in A : \text{condition}\}`
$\|\,\|$	7, 14, 17, 21, 43	`| |`
$\ldots$	5	`\ldots`
$[n]$	5, 28	`[n]`
$[a, b], [a, b)$ etc.	16	`[a,b], [a,b)`
$[a, b[,]a, b],]a, b[$	16	`[a,b[, ]a,b], ]a,b[`

---

[2]LaTeX is a document preparation system that is excellent for writing mathematics. Various packages may be required to produce these symbols such as amsfonts, amsmath, and eufrak.

[3]Requires the undertilde package.

[4]Requires the harpoon package.

Symbol	Page(s)	LaTeX
(n, m)	32	`(n,m)`
$\lfloor\,\rfloor$	28	`\lfloor \rfloor`
$\lceil\,\rceil$	28	`\lceil \rceil`
$\|\,\|$	40	`| |`
$\|\,\|_p, \|\,\|_\infty$	40	`| |_p, | |_\infty`
$\|\,\|, \|\,\|_p$	45	See discussion on page 45.
$\|\,\|_F$	45	`| |_F`
$\langle\,\rangle$	40, 57	`\langle \rangle`
$\langle\,\vert\,\rangle$	40	`\langle	\rangle`
$\langle x\vert, \vert y\rangle$	40	`\langle x |,	y \rangle`
$[\mathbf{a}, \mathbf{b}, \mathbf{c}]$	41	`[\mathbf{a},\mathbf{b},\mathbf{c}]`
$[\![\;]\!]$	35	`\llbracket \rrbracket`
Relations		
$\in$	5	`\in`
$\notin$	5	`\notin`
$\ni$	5	`\ni`
$\subseteq, \subset$	6	`\subseteq,\subset`
$\supseteq, \supset$	6	`\supseteq,\supset`
$:=, \triangleq, \overset{\text{def}}{=}$	15	`:=, \triangleq,` `\buildrel{\text{def}\over=}`
$\equiv$	15, 32, 34	`\equiv`
$\parallel$	22	`\parallel`
$\perp$	22, 40	`\perp`
$a\vert b$	32	`a|b`
$\approx$	63	`\approx`
$\cong$	22	`\cong`
$\sim$	22, 63	`\sim`
$\succeq, \succ, \preceq, \prec$	45	`\succeq, \succ, \preceq, \prec`
$\propto$	63	`\propto`
$\asymp$	64	`\asymp`
$\ll, \gg$	66	`\ll, \gg`
Operations		
$+, -$	18	`+,-`
$*$	18, 55	`*`
$\div, /, \frac{a}{b}$	18	`\div, /, \frac{a}{b}`
$\cdot$	18, 40	`\cdot`
$\times$	7, 18, 41	`\times`
$\sqrt{x}, \sqrt[n]{x}$	27	`\sqrt{x},\sqrt[n]{x}`
$B^A, **$	7, 18	`B^A, **`
$\cup, \cap$	6	`\cup, \cap`
$\setminus$	7	`\setminus`
$\wedge, \vee$	8, 11, 18	`\wedge, \vee`

Chart 71

Symbol	Page(s)	LaTeX
&	11	\&
$\underline{\vee}, \oplus$	11	\veebar, \oplus
$\overline{\wedge}$	11	\barwedge
$\neg, \sim$	11	\lnot, \sim
$\pm$	14, 63	\pm
$\mp$	15	\mp
$\circ$	26	\circ
$\otimes, \oplus$	44	\otimes, \oplus
$x^{\overline{k}}, x^{\underline{k}}$	31	x^{\overline k}, x^{\underline k}
$\binom{n}{k}$	31	\binom{n}{k}
$\binom{n}{k_1\ k_2\ \cdots\ k_t}$	31	\binom{n}{k_1\ k_2\ \cdots\ k_t}
$\left(\!\binom{n}{k}\!\right)$	32	\left(\!\binom{n}{k}\!\right)
$\binom{n}{k}_q$	32	{\binom{n}{k}}_q
$\left(\frac{a}{p}\right), (a\|b)$	33	\left(\frac{a}{p}\right), (a|b)
$\genfrac{[}{]}{0pt}{}{n}{k}$	32	\genfrac{[}{]}{0pt}{}{n}{k}
$\genfrac{\{}{\}}{0pt}{}{n}{k}$	32	\genfrac{\{}{\}}{0pt}{}{n}{k}
Arrows		
$\Rightarrow$	11	\Rightarrow
$\Leftarrow$	11	\Leftarrow
$\Leftrightarrow, \Longleftrightarrow$	11	\Leftrightarrow, \iff
$\Rightarrow\!\Leftarrow$	11	\Rightarrow\Leftarrow
$f : A \to B$	25	f:A \to B
$f : A \hookrightarrow B$	35	f:A \hookrightarrow B
$f : A \twoheadrightarrow B$	35	f:A \twoheadrightarrow B
$\mapsto, \buildrel{f}\over{\mapsto}$	25	\mapsto, \buildrel{{f}\over{\mapsto}}
$\uparrow, \downarrow$	49	\uparrow,\downarrow
$\buildrel{d}\over{\longrightarrow}$	61	\buildrel{d}\over{\longrightarrow}
$\buildrel{\mathcal{D}}\over{\longrightarrow}$	61	\buildrel{\mathcal{D}}\over {\longrightarrow}
$\buildrel{\mathcal{L}}\over{\longrightarrow}$	61	\buildrel{\mathcal{L}}\over {\longrightarrow}
$\buildrel{P}\over{\longrightarrow}$	61	\buildrel{P}\over {\longrightarrow}
$\buildrel{a.s.}\over{\longrightarrow}$	61	\buildrel{\text{a.s.}}\over {\longrightarrow}

Symbol	Page(s)	LaTeX
Calculus		
$\frac{df}{dx}, \frac{d^2f}{dx^2}$	50	`\frac{df}{dx}, \frac{d^2f}{dx^2}`
$\frac{d}{dx}, \frac{d^n}{dx^n}$	50	`\frac{d}{dx}, \frac{d^n}{dx}^n`
$\left.\frac{df}{dx}\right\|_{x=a}$	25, 50	`\left.\frac{df}{dx}\right|_{x=a}`
$\dot{y}, \ddot{y}$	50	`\dot y, \ddot y`
$\frac{\partial f}{\partial x_j}$	50	`\frac{\partial f}{\partial x_j}`
∂D	55	`\partial D`
f_x, f_y	50	`f_x, f_y`
$D_x f, D_\mathbf{v} f$	50	`D_x f, D_{\mathbf{v}} f`
$\mathbf{D}\mathbf{f}$	52	`D \mathbf{f}`
$\frac{\partial(u,v,w)}{\partial(x,y,z)}$	53	`\frac{\partial(u,v,w)}` `{\partial(x,y,z)}`
$\frac{\partial^2 f}{\partial x_i \partial x_j}$	50	`\frac{\partial^2 f}` `{\partial x_i \partial x_j}`
f_{xx}, f_{xy}	50	`f_{xx}, f_{xy}`
$f_{i,j}, f_{,j}$	52	`f_{i,j}, f_{,j}`
∇f	51	`\nabla f`
Δf	51	`\Delta f`
$\nabla \times \mathbf{f}$	53	`\nabla\times\mathbf{f}`
$\nabla \cdot \mathbf{f}$	53	`\nabla\cdot\mathbf{f}`
$\int f(x)\,dx$	54	`\int f(x) \, dx`
$F(x)\Big\|_a^b$	54	`F(x)\Bigr|_a^b`
$\iint f(x,y)\,dxdy$	54	`\iint f(x,y)\, dx dy`
$\int_A, \int_\gamma$	54	`\int_A, \int_\gamma`
$\oint, \oint_{\partial D}$	55	`\oint, \oint_{\partial D}`
$f * g$	55	`f*g`
Other		
$\emptyset, \varnothing$	5	`\emptyset,\varnothing`
#	7	`\#`
$\therefore, \because$	11	`\therefore, \because`
$\Box, \blacksquare$	11	`\Box or \qed, \blacksquare`
☺	12	`\smiley`[5]
$\angle$	21	`\angle`
$\triangle ABC$	21	`\triangle ABC`
$f(x) \in O(g(x))$	65	`f(x)\in O(g(x))`
${}_2F_1\left(\begin{smallmatrix}a,b\\c\end{smallmatrix}\middle\|x\right)$	29	LaTeX code shown on page 29
$R[x], R[\![x]\!]$	35	`R[x], R\llbracket x \rrbracket`
$\aleph_0$	20	`\aleph_0`

---

[5]Requires the wasysym package.

Alphabetical Notation Index

Use this index to find notation that is based on a Latin letter (e.g., $\mathcal{L}$ or $\mathbb{R}^n$) or a word formed by Latin letters (e.g., det). Use the Greek index (page 77) to find notation based on Greek letters and use the Chart (page 69) to find other notation. Finally, use the Topic Index (page 79) to find concepts by keyword.

Greek Notation Index

Use this index to find notation that is based on Greek letters. See also Figure 1.1 (on page 2) for a table of the Greek alphabet in its entirety.

Topic Index

Use this index to find specific topics. To find notation based on Latin letters, use the Alphabetical Notation Index (page 73). To find notation based on Greek letters, use the Greek Index (page 77). To find other notation, use the Chart (page 69).

Made in United States
Troutdale, OR
11/25/2024

25268688R00053